UNDERSTANDING AND PREVENTING FALLS

T0174207

UNDERSTANDING AND PREVENTING FALLS

EDITED BY
Roger Haslam
David Stubbs

CRC Press
Taylor & Francis Group
Boca Raton London New York

CRC Press is an imprint of the
Taylor & Francis Group, an **informa** business
A TAYLOR & FRANCIS BOOK

CRC Press
Taylor & Francis Group
6000 Broken Sound Parkway NW, Suite 300
Boca Raton, FL 33487-2742

First issued in paperback 2019

ISBN-13: 978-0-415-25636-0 (hbk)
ISBN-13: 978-0-367-39185-0 (pbk)
Library of Congress Card Number 2005050528

Library of Congress Cataloging-in-Publication Data

Understanding and preventing falls / edited by Roger Haslam and David Stubbs.
     p. cm.
  Includes bibliographical references and index.
  ISBN 0-415-25636-4 (alk. paper)
    1. Falls (Accidents)--Prevention. 2. Falls (Accidents) in old age--Prevention. I. Haslam, Roger. II. Stubbs, David, 1947-

RC952.5.U535 2005
613.6--dc22   2005050528

Preface

Falls are a leading cause of injury in the workplace, health care, home, and other domestic situations, to the extent that it might reasonably be argued that the falls problem is of international epidemic proportions. Although reductions have occurred in the number and severity of injuries from other causes over recent decades, the incidence of injuries from falling has remained at a consistently high level. Despite this, many falls are preventable. This book provides an authoritative guide to the nature and extent of the falls problem, emphasising that falls occur due to a combination of factors, such as the design and condition of the walking surface, footwear, lighting, and weather conditions. These environmental factors interact with the health, fitness, strength, balance, vision, and activities of individuals involved. Falls in differing circumstances, including slips and trips on the level, falls on stairs, falls among older people, falls in the workplace, and falls during entry/egress from vehicles, are considered in terms of their causation and prevention, drawing on the latest research. The case is made for a systems approach to falls prevention, taking into account the complex interaction between individuals and the environment. Although more obvious concerns, such as the nature of footwear and the walking surface, are important, prevention of falls also requires attention to issues such as knowledge of falls risks and corresponding behaviour, together with influences, such as those affecting job and task design and the implementation of effective risk management procedures.

The Editors

Roger Haslam, Ph.D., is Professor of Ergonomics, Director of the Health and Safety Ergonomics Unit, and head of the Department of Human Sciences at Loughborough University. Professor Haslam's research addresses human factors aspects of health and safety, and he has a long-standing interest in falls in both occupational and domestic situations. Research in this area has included consideration of falls not only in distribution, food manufacturing, catering, and forestry industries, but also in sporting facilities and among older people in and around the home. Professor Haslam is an editor of the scientific journal *Applied Ergonomics* and he was until recently Chairman of Council of the Ergonomics Society.

David Stubbs, Ph.D., is Professor of Ergonomics at the Robens Centre for Health Ergonomics, EIHMS, at the University of Surrey. He is the former Director of the Robens Institute of Industrial and Environmental Health and Safety at Surrey and past President of the Ergonomics Society. His interest in slip, trip, and fall accidents (STFA) dates back to 1982 when the Robens Institute and the Medical Commission on Accident Prevention jointly ran the first of five international STFA conferences. His main research interests relate to the ageing workforce, and most recently, his attention has been focussed on systems design, accidents, and incidents in health-care settings.

Contributors

Liz Ashby
Centre for Human Factors and
  Ergonomics
Rotorua, New Zealand

Tim Bentley
Massey University, Albany Campus
Auckland, New Zealand

Katherine Brooke-Wavell
Loughborough University
Leicestershire, United Kingdom

Fadi Fathallah
University of California — Davis
Davis, California

Raoul Grönqvist
Finnish Institute of Occupational
  Health
Helsinki, Finland

Roger Haslam
Loughborough University
Leicestershire, United Kingdom

Denise Hill
Loughborough University
Leicestershire, United Kingdom

Peter Howarth
Loughborough University
Leicestershire, United Kingdom

Paul Lehane
London Borough of Bromley
Bromley, United Kingdom

Paul Lemon
Health and Safety Laboratory
Buxton, United Kingdom

Stephen Lord
Prince of Wales Medical Research
  Institute
Randwick, New South Wales,
  Australia

Hylton Menz
La Trobe University
Victoria, Australia

Richard Parker
Centre for Human Factors and
  Ergonomics
Rotorua, New Zealand

Mike Roys
Building Research Establishment
Garston, Watford, United Kingdom

Catherine Sherrington
Prince of Wales Medical Research
  Institute
Randwick, New South Wales,
  Australia

Joanne Sloane
Loughborough University
Leicestershire, United Kingdom

David Stubbs
Surrey University
Guildford, United Kingdom

Stephen Taylor
Health and Safety Executive
Bootle, Merseyside

Steve Thorpe
Health and Safety Laboratory
Buxton, United Kingdom

Contents

Chapter 1

Introduction

Roger Haslam and David Stubbs

CONTENTS

1.1 Background

Falls are an intriguing problem. On the one hand, they are commonplace events, afflicting the human species from childhood through to old age. Often, the outcome is no more serious than a loss of dignity and a degree of embarrassment for the individual concerned. When injuries result, however, they can be debilitating and far-reaching, with detrimental consequences for the injured person's family, his or her colleagues, and employer. Set against this, efforts aimed at prevention face the considerable challenge of foiling the many, varied, and interacting circumstances that cause falls to occur.

The scale of the falls epidemic was highlighted almost 25 years ago at the first University of Surrey "Slipping, Tripping and Falling Accidents" (STFA) conference in 1982. Two years later, at the second STFA '84 meeting, Lord Porrit, introducing the event, was perplexed by the limited attention fall accident research was receiving, disproportionate to the " ... gargantuan world-wide epidemic" of the problem (Porritt 1985). Porritt went on to observe that the wider community appears to form " ... a vast, disinterested host of young and old who fall over at work and at play, in the home and in the streets, seemingly without caring, except perhaps at the time of the individual incident." Despite these strong comments, it has taken many years for this persistent leading cause of injury at home and in the workplace to be taken seriously. Fortunately, within the last 10 years, a widespread, international effort has been directed at improving understanding of the causes of falls and their prevention. Much still needs to be done, however, toward achieving a meaningful, sustained reduction in the high incidence of fatalities and injuries arising from falls.

It is appropriate to comment on what constitutes a fall. An early definition, from a study in a geriatrics context, defined a fall as "an untoward event in which the patient comes to rest unintentionally on the floor" (Morris and Isaacs 1980). This description is worthy of remark, though, because it would seem to include incidents such as an individual being knocked down after being struck violently by another person, while excluding a situation where someone falls onto something other than the floor (e.g., an item of furniture). In both cases this seems counterintuitive. The Kellogg International Working Group on the Prevention of Falls by the Elderly (1987) addressed these points to some extent by defining a fall as "an unintentional event that results in a person coming to rest on the ground or another lower level." For the purposes of the present book, the Kellogg Working Group definition is a useful starting point. It can be added to this that fall incidents will usually involve a person moving about his or her environment or, when stationary, having his or her balance disrupted through movement of the surface on which they are standing (e.g., as might occur when standing on a moving bus or train). Other circumstances, such as the collapse of a person due to a medical condition, such as epilepsy or syncope, are not excluded from consideration, but should be represented as discrete categories of falling, with distinct causation. The same applies to an individual falling as a consequence of being struck by an object. A person losing his or her balance and falling as a result of being jostled in a moving crowd, for example, would appear to be of legitimate interest from a fall prevention perspective.

On occasions in the following chapters, falls are referred to as accidents. The term "accident" can be controversial in the field of injury prevention, owing to the possible interpretation that events leading to injury, described in this way, are due largely to chance and cannot be averted (Evans 2001). The term "accident" remains in widespread use among the safety community, however, and is used comfortably in this book in connection with falls and falling. In commenting on this point, we can be explicit that as falls are clearly more prevalent in particular situations and among certain groups in the population than others, falls cannot be regarded as random events. Moreover, it is a fundamental premise of this book that falls are not inevitable and can be prevented.

1.1.1 *Extent of the Falls Problem*

Falls happen on the level, on slopes, on steps and stairs, and from height, with differing causes and consequences. Together, they form a major source of injury, imposing a substantial social and economic burden on society. Statistics for the United Kingdom indicate that falls were responsible for 22% of accidental deaths in 2002 across all age groups (RoSPA 2004). In the workplace in Great Britain, falls are responsible for at least one in five deaths and one in three reported nonfatal injuries among employees (Figure 1.1) (HSE 2004). The Health & Safety Executive (HSE 2003) has estimated that slips and trips cost employers £512 million each year in lost production and other expense, with a cost to the health services of £133 million per year. In the home, U.K. Home Accident Surveillance System (HASS) data for 2002 suggest that falls account for almost half (46%) of all home accidents, with an annual incidence rate estimated at 2108 per 100,000 of the population (Figure 1.2) (DTI 2003).

Although international comparisons are impeded by different reporting and recording procedures, data from the U.S. Bureau of Labor Statistics (2004) indicate that falls are involved in 12% of fatal occupational accidents. Similarly, data presented by the National Safety Council (2003) lists falls as the third leading cause of unintentional injury deaths in the general population, behind motor vehicle and poisoning related fatalities, accounting for 15% of the total. Studies in the United States examining the economic impact of falls among older people have estimated the total cost of all fall injuries for people age 65 or older in 1994 to be $20.2 billion (Englander et al. 1996), with an average health care cost of a fall injury for individuals age 72 and over of $19,440 (including hospital, nursing home, emergency room, and home health care, but not physician services) (Rizzo et al. 1998). A similar pattern exists in other countries: Falls are a major cause of unintentional death and injury worldwide.

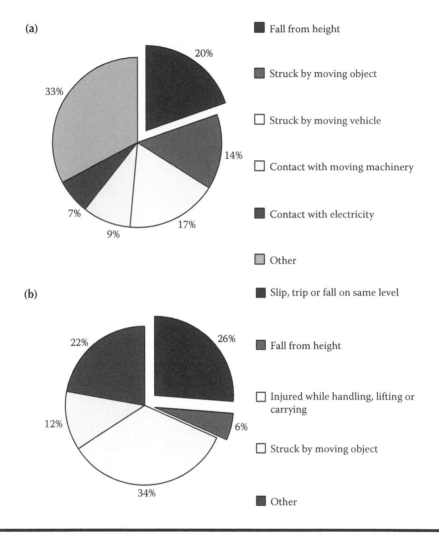

Figure 1.1 **(a) Fatal occupational injuries to employees in Great Britain 2002/03 (HSE, 2004). (b) All reported occupational injuries to employees in Great Britain 2002/03 (HSE, 2004).**

1.1.2 Types of Falls

In practice, falls are categorised in several ways, according to the circumstances of the fall, age of the person falling and the prevention and research communities most directly addressing the problem (Figure 1.3). In terms of location, it is customary to differentiate between falls on the level, falls on steps and stairs, and falls from height, including those from ladders, raised walking areas, windows, and balconies. Falls in any of

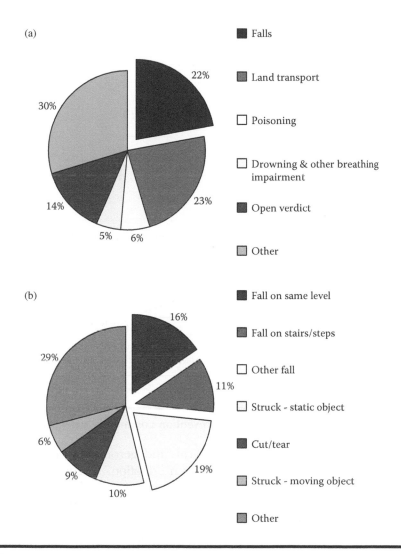

(a)

■ Falls

■ Land transport

□ Poisoning

□ Drowning & other breathing impairment

■ Open verdict

▨ Other

(b)

▨ Fall on same level

▨ Fall on stairs/steps

□ Other fall

□ Struck - static object

■ Cut/tear

▨ Struck - moving object

▨ Other

Figure 1.2 **(a) Accidental deaths in the U.K. 2002 (RoSPA, 2004). (b) Home accidents in the U.K. 2002 (DTI, 2003).**

these locations may involve a slip or a trip, with these being the common antecedents of falls on the level. Differentiation can be made between falls occurring among children, healthy adults and older people, in terms of individual capability, limitations and the nature of activities typically involved. Differences also exist in the circumstances surrounding falls that happen in domestic, leisure, sporting, and work situations.

An interesting separation exists between authorities and prevention communities addressing falls among different age groups and falls in and away from the workplace. Falls among older people, for example, are a

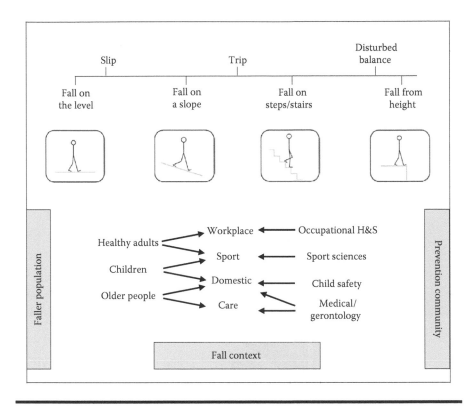

Figure 1.3 Primary fall, faller, and prevention community categorisations.

priority concern for healthcare, medical, and gerontology professionals, and have received considerable research attention. Occupational health and safety professions have primary responsibility for falls in the workplace, where there has perhaps been less research, certainly in terms of the effectiveness of intervention. Other important stakeholders are designers and architects concerned with the built environment. Given the generic aspects of falls, wherever or to whom they occur, it is perhaps surprising how little interaction there has been between specialists from different fields, with little communication or sharing of experience. Underlying this, presumably, is that the development of safety prevention activity and research is an evolutionary process, depending on where expertise is located at the outset, how this develops and the subsequent influence of funding and regulatory bodies.

1.1.3 Causes of Falls

In the most general sense, falls involve a loss of balance due to some reason, which results in a person falling to the ground or other lower level. As with other accidents, the events leading to a fall are frequently multifactorial, involving an interaction of conditions, the combination of which is necessary for the incident to occur. Prevention of falls requires a commensurate multifaceted approach.

Aspects of the environment involved in falls are the foot–floor interface and the presence of trip hazards. The frictional characteristics of footwear and flooring materials affect the likelihood of slipping, with these influenced by their condition and maintenance and also the presence of contaminants (e.g., water, ice, powders). Obstacles in the walkway may lead to tripping if they go undetected. Because the clearance between feet and the floor is so small during normal gait, deviations in the walking surface of as little as 10 mm may be sufficient to cause a trip.

Features of stairs affecting safety include the stairway dimensions (step height and depth), conspicuity of step edges, and provision of handrails. The stability of ladders depends on the angle at which they are placed and whether they have been secured or tied in at their top. When at height, the stability of the platform on which individuals are positioned is important, together with the provision of guardrails. The height of guard and balcony rails in relation to a person's centre of gravity affects the protection afforded.

Falls from height, although not considered in detail by this book, are the types of fall responsible for the most serious injuries, especially in the building and construction industry. Inappropriate use of unsecured ladders, unguarded edges or openings to a lower level, working at height without use of suitable supporting platforms (scaffolding), and entering onto fragile roofs without crawler boards are typical causes of such falls.

Personal characteristics involved in fall accidents include gait, balance, stature, strength, vision, and behaviour (Figure 1.4). These, in turn, are influenced by individual health, fatigue, use of medication, alcohol, features of the environment (e.g., lighting and floor surface), and a person's activity (e.g., load carriage or performance of a cognitive task).

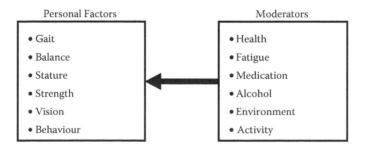

Personal Factors	Moderators
• Gait	• Health
• Balance	• Fatigue
• Stature	• Medication
• Strength	• Alcohol
• Vision	• Environment
• Behaviour	• Activity

Figure 1.4   Personal factors in falls.

Human ability to walk and balance is a complex process, involving integration of inputs from the visual, vestibular, and somatosensory systems. Stature influences the location of an individual's centre of gravity and inherent stability. Strength affects ability to recover from disturbances to balance, and use of supports such as handrails. Vision, in addition to its contribution to balance, also enables monitoring of the walking surface and detection of obstacles. Behaviour may vary in terms of the caution that individuals exercise in different circumstances and the nature of the environments people establish around themselves and then seek to negotiate.

1.1.4 Fall Outcomes

Where physical harm results from a fall on the level, injuries may involve lower or upper limb fractures or sprains, back or head injuries, and bruising or contusions. Falls on the level are less likely to lead to serious injury and, if fortunate, the individuals involved may be able to return to their feet, brush themselves down, and continue on their way. In a small number of instances, falls on the level are fatal, where head injury occurs or where a fall exposes the victim to another hazard (e.g., approaching motor vehicle).

Falls from height are often serious in their outcome. The extent of the forces generated and severity of resulting injuries depends on the distance through which the victim falls, any features of the environment with which contact occurs during falling and the nature of the surface on which the body lands. Rigid, inelastic surfaces will cause most damage to the body, whereas a surface capable of yielding to an extent will absorb some of the energy on impact. On some occasions, a falling person may be unfortunate enough to be impaled on a protruding object, causing internal injuries.

In addition to physical injury, individuals may experience psychological harm, with distress caused by the event itself and subsequent pain, immobilisation, and inability to perform normal activities of daily living. This, in turn, may lead to work, financial, and relationship problems, accompanied by depression or anxiety. In some cases, a continuing fear of falling persists, leading to a diminished quality of life for afflicted individuals.

1.2 Book Structure

This book is divided into two sections. The first part examines the current state of knowledge relevant to understanding the causes and prevention of falls, with chapters covering human ambulation and balance on the level and on steps and stairs, followed by chapters looking at the involvement of vision and ageing. The second half of the book presents a series of case studies, illustrating how falls arise in different circumstances and differing approaches to prevention.

As indicated previously, certain aspects of falls and falling have received more attention than others. Tribological concerns (i.e., frictional properties at the foot–floor interface), for example, have been the focus of research effort, spanning at least two decades (Chang et al. 2003). The last few years have also seen appreciable growth in the interest given to falls among older people, with research and intervention programmes now in place in many countries. This is demonstrated by several substantive reviews that have now been published (e.g., American Geriatrics Society et al. 2001; Lord et al. 2001; Gillespie et al. 2003). Four chapters in the book are therefore devoted to these issues.

This book does not give explicit attention, beyond the brief mention in this introductory chapter, to certain categories of falls and fallers, namely falls among infants and children and falls from height. The editorial decision to confine attention to falls among adults on the level and on steps and stairs has been in large part due to the maturity of the research evidence available for different categories of falls and fallers.

Following this introductory chapter, Raoul Grönqvist (Chapter 2) provides a detailed account of the characteristics of human gait and foot–floor interaction, from a frictional, tribology perspective. Grönqvist explains that falls are the outcome of an unrecoverable perturbation of balance, due to a slip, trip, or other cause. Understanding the complexity of how we move and the manner in which a slip or trip may occur are central to understanding falls and how they might be prevented. The chapter proceeds to discuss the friction that occurs between the feet and the floor and how the performance of this can be affected by different footwear and flooring characteristics. Consideration is then given to the mechanical and human-based approaches to assessing these.

Mike Roys, Chapter 3, discusses the circumstances surrounding falls on steps and stairs, steps and stairs being common features of our everyday environment. Human gait on stairs is described, followed by examination of how this can be disrupted. Aside from more obvious problems, such as objects left on stairs, Roys explains how even small variations in step dimensions can be sufficient to cause a trip or misstep. A surprising point is that many step treads found in the built environment are too small for the majority of adult feet, and Roys suggests that this may be a source of problems for more vulnerable users. The importance of appropriately designed and positioned handrails is discussed, along with the need for satisfactory lighting.

Peter Howarth, a qualified ophthalmologist, explains in Chapter 4 the important role that is played by vision in traversing our surroundings. One aspect is the contribution vision makes in allowing us to monitor the environment as we move, detecting changes in flooring and levels, as well as obstacles and slip hazards. Vision also operates in the maintenance of balance, augmenting our vestibular and somatosensory systems. The characteristics of our visual system affect the requirements for lighting and considerations such as conspicuity (i.e., the extent to which objects stand out from their background). These have an effect on our ability to notice objects in our walking path or distinguish step edges on stairs, for example. As with many human attributes, vision varies from one person to another and as we age. The processes involved in this are described, along with the problems that may accompany deteriorating vision from a falls perspective and possible unwanted consequences from the wearing of spectacles.

In Chapter 5, Stephen Lord and his co-authors give a detailed overview of the causes of falls and approaches to their prevention among older people. Prevalence of falls increases significantly in later life, with at least one in three individuals age 65 and over experiencing a fall each year, with the rate increasing to one in two for those over age 80. Increasing vulnerability arises due to declining physical and sensory capabilities. The health problems that accompany ageing and the effects of medication used to treat these are also factors. When older people experience injury as a result of falling, healing and recuperation may take longer than for their younger counterparts. In terms of prevention, studies of targeted interventions have found measures that increase individual capability, for example through muscle strengthening, balance training, and medication review, to be effective in reducing numbers of falls. Multifactorial health and environmental fall prevention programmes have also been found to be beneficial.

Chapters 6 and 7 consider differing approaches to investigating falls, first with Tim Bentley and Roger Haslam (Chapter 6) examining research methods for understanding falls at the population level, drawing on their work examining fall causation among postal delivery workers. Chapter 6 discusses archival data analysis, such as information held in occupational injury databases, including analysis of narrative text fields, along with other research approaches that may be used to supplement this. Investigations independent of individual fall incidents can be useful in gathering background information and stakeholder opinion on the factors involved in falls in a given situation. Incident specific follow-up investigations can be a resource intensive method, but can be useful in providing valuable evidence to confirm findings from other analyses.

Paul Lehane and David Stubbs (Chapter 7) then go on to consider investigation of individual fall incidents, within the scenario of workplace managers and supervisors executing their responsibilities. This routine investigative activity might be expected to make an important contribution to fall prevention, providing learning and understanding for influential front line managers in fall causation. This chapter brings a novel perspective to the problem of falls, focussing on nonspecialist perceptions of falls, important when it comes to attempting to manage fall risk.

In Chapter 8, Steve Thorpe, Paul Lemon, and Stephen Taylor, review the efforts being made to reduce the incidence of fall-related injuries in the United Kingdom, through the assessment and prevention of pedestrian slipping. Many techniques and instruments have been proposed for the assessment of slipperiness, with vested commercial interests making this a controversial area. Thorpe et al. describe the method recommended by HSE, the regulatory body for workplace health and safety in Great Britain. Attention to the "designers dilemma" identifies the complexity of specifying flooring, with the conclusion that there is considerable scope to provide safer flooring in public areas and the workplace. The chapter then presents two practical examples of problem flooring installations and discusses the issues surrounding these.

Tim Bentley, Richard Parker, and Liz Ashby (Chapter 9) describe, as a case study, an epidemiological analysis of injury data from the forestry industry in New Zealand. Characterised by outdoor working in the roughest of terrain, falls are a significant problem in forestry operations. The benefits of undertaking an analysis of the nature presented in this chapter is in providing an objective basis for prioritising issues for further investigation or intervention. Bentley et al. conclude their chapter with an overview of a field trial addressing one of the priorities: slips from logs and on other forestry debris. The evaluation of spiked boots, which examined effects on worker safety, productivity, and workload, concluded that the spiked footwear reduced the risk of falling, without any detrimental effects on other aspects of performance.

Another situation resulting in fall injuries is entry and egress from commercial vehicles. In a further case study, Fadi Fathallah (Chapter 10) describes how falls arise due to individuals having to climb up or down from a driving cab or trailer deck, often with absent or poorly designed hand–foot support. Even where entry/egress aids are provided, workers often fail to use them, preferring to jump down onto the ground. Fathallah includes an overview of an experimental study, which examined the slip potential and forces involved for different vehicle configuration and egress methods. This study demonstrated why fall injuries occur during unaided egress from vehicles, indicating the impact forces and high slip–fall potential present in these circumstances. Prevention of falls from vehicles needs attention to the vehicle design, with provision of hand–foot support that takes account of user anthropometry, combined with worker education and training, alongside attention to work organisation aimed at reducing the frequency with which workers have to enter and leave their vehicles to perform tasks.

Chapter 11 examines the manner in which behaviour affects the safety of older people on stairs. Although the intrinsic personal and extrinsic environmental factors in falls among older people have been the subject of some considerable attention, the influence of behaviour is less well understood. Behaviour operates at several levels in affecting risk of falling. First, the manner in which individuals interact directly with their environment has a direct bearing on safety. Hurrying on stairs while carrying something bulky is an undesirable combination in this respect. In addition to this, individual actions and decisions affect the safety of the environment that has to be negotiated, through the introduction of slip and trip hazards or the provision of adequate lighting, for example. Behaviour also affects individual capability, with the use of alcohol, medication, use or nonuse of spectacles, and efforts to maintain physical fitness all being issues for older people. The stairs case study presented in Chapter 11 identifies a pattern of behavioural influences, many of which are also relevant to falls among younger adults and in circumstances other than the use of stairs.

Finally, the concluding chapter (Chapter 12) summarises the wide-ranging causes of falls and approaches to their prevention.

References

American Geriatrics Society, British Geriatrics Society and American Academy of Orthopaedic Surgeons Panel on Falls Prevention, 2001, Guideline for the prevention of falls in older persons. *Journal of the American Geriatrics Society*, 49, 664–672.

Bureau of Labor Statistics, 2004, National census of fatal occupational injuries in 2003. *Bureau of Labor Statistics News*, September 22, 2004.

Chang W.-R., Courtney T. K., Grönqvist R., and Redfern M. S., 2003, *Measuring slipperiness: human locomotion and surface factors* (Taylor & Francis: London).

Department of Trade and Industry (DTI), 2003, *24th final report of the home and leisure accident surveillance system* (DTI: London).

Englander F., Hodson T. J., and Terregrossa R. A., 1996, Economic dimensions of slip and fall injuries. *Journal of Forensic Science*, 41, 733–746.

Evans S. A., 2001, Banning the "A word": where's the evidence? *Injury Prevention*, 7, 172–175.

Gillespie L. D., Gillespie W. J., Robertson M. C., Lamb S. E., Cumming R. G., and Rowe B. H., 2003, Interventions for preventing falls in elderly people (Cochrane Review). In: *The Cochrane Library*, Issue 4 (Wiley: Chichester).

Health & Safety Executive (HSE), 2004, *Health and safety statistics highlights 2003/04* (Health and Safety Executive: Merseyside).

Health & Safety Executive (HSE), 2003, *Preventing slips and trips and work* (Health and Safety Executive: Merseyside), INDG225(rev1).

Kellogg International Working Group on the Prevention of Falls by the Elderly, 1987, The prevention of falls in later life. *Danish Medical Bulletin*, 34, 215–219.

Lord S., Sherrington C., and Menz H. B., 2001, *Falls in older people: risk factors and strategies for prevention* (Cambridge University Press: Cambridge).

Morris E. V. and Isaacs B., 1980, The prevention of falls in a geriatric hospital. *Age and Ageing*, 9, 181–185.

National Safety Council, 2003, *Report on injuries in America, 2002* (National Safety Council: Itasca, Illinois).

Porritt, the Right Honourable Lord, 1985, Slipping, tripping and falling: familiarity breeds contempt. *Ergonomics*, 28, 947–948.

Rizzo J. A., Friedkin R., Williams C. S., Nabors J., Acampora D., and Tinetti M. E., 1998, Health care utilization and costs in a Medicare population by fall status. *Medical Care*, 36, 1174–1188.

Royal Society for the Prevention of Accidents (RoSPA), 2004, *Accidental deaths in the U.K. 2002* (RoSPA: Birmingham).

Chapter 2

Walking on the Level: Footwear and the Walking Surface

Raoul Grönqvist

CONTENTS

2.1 Introduction

The majority of our time as pedestrians is spent negotiating level walking conditions, with slipping or tripping an important cause of falls in these circumstances. Whereas intrinsic or individual risk factors appear to be a primary cause of falls in older people (Waller 1978), extrinsic or environmental factors probably initiate most falls among the healthy adult population (Grönqvist 1995a). The latter includes problems with the walking surface (e.g., contamination, irregularities, or obstacles), deficient footwear or inadequate illumination (Grönqvist et al. 2001c; Redfern et al. 2001). Walkway surface contaminants (e.g., water, oil, organic debris, soil, snow) may be involved as contributing factors in as many as 80% of slip-related injury incidents (Grönqvist 1999; Leclercq 1999).

This chapter considers fall prevention for walking on the level, from the perspective of shoes and walkways. A prerequisite to this is to understand human locomotion and the interaction at the foot–floor interface. Previous research is reviewed, with discussion of gait and balance, fall risk on the level, biomechanics of slips, trips and falls, postural strategies for the recovery of stability, methods for the measurement of slipperiness, implications for falling, and guidance for prevention.

2.2 Walking on the Level

This section describes normal gait and the situations in which balance may be challenged.

2.2.1 Normal Gait and Balance

Even normal gait on a safe surface can be regarded as inherently instable, because equilibrium is lost and regained with every step (Carlsöö 1962): "Equilibrium is lost with the take-off of the propelling foot, when the body's centre of gravity momentarily lies beyond the anterior border of the supporting surface, and regained as soon as the swinging leg is extended forward and its heel touches the ground." This instability of upright posture is exploited in walking to assist in propelling the body forward.

During locomotion, a dynamic interplay occurs between sensory systems (vision, vestibular organ, proprioception, and pressoreceptors), which control posture, timing and placement of successive steps (step length and cycle time), on the one hand, and surface-related (friction and wear) phenomena between shoes and walkways on the other (Grönqvist et al. 2001c). The manner in which balance is maintained and the strategies applied (at ankle, hip or combined) for coordinating legs and trunk to maintain the body in equilibrium with respect to gravity are contingent on the human capacity to anticipate postural demands (Nashner 1985; Winter 1995). Gait stability depends on the ability to move and control the body's centre of gravity or mass (COM) back to a position over the base of support (BOS) and the centre of foot support or pressure (COP) (Figure 2.1). The alignment of body segments over the BOS must be kept such that the projection of the body COM falls within the boundaries of a "stability region" (Grönqvist et al. 2001d).

The region of stability can be predicted based on the physical constraints of muscle strength, size of BOS, and floor surface contact forces within an environment (Pai and Patton 1997). The size and shape of this region also depends on the mechanical properties of the body and the response latencies of the neuromuscular system for postural control (Redfern and Schuman 1994). The position of the body's COM must be controlled within the limits of stability during the stance phase (double-leg support) and the swing phase (one-leg support) by continually establishing a new BOS. In summary, balance and stability in walking involve controlling movements of the body's COM and the placement of the BOS.

2.2.2 Kinematics and Ground Reaction Forces

2.2.2.1 Foot Trajectory and Heel Landing

The foot trajectory needs to be controlled during the gait cycle for safe ground clearance, accompanied by a gentle heel or toe landing (Winter 1991). To allow sufficient ground clearance during the swing phase, the

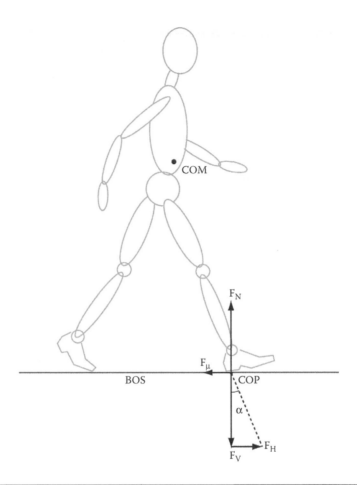

Figure 2.1 Minimum friction coefficient for slip avoidance based on the equilibrium of forces at heel contact: $\mu_{min} = F\mu/F_N = F_H/F_V$. $F\mu$ = **frictional force,** F_N = **normal force,** F_H = **horizontal ground reaction force,** F_V = **vertical ground reaction force, BOS = base of support, COM = centre of body mass, and COP = centre of foot pressure. (From Grönqvist, R., Chang, W.R., Courtney, T.K., Leamon, T.B., and Strandberg, L., 2001c, *Ergonomics*, 44, 1102–1117.)**

leg is flexed at the knee and dorsiflexed at the foot and toes. Ground clearance is typically around 6 mm, with irregularities of as little as 10 mm in the height of the walking path sufficient to cause a trip. A gentle heel landing reduces collision forces during weight acceptance (Cappozzo 1991), desirable for minimizing "required" and "utilized" friction (see below).

Force interactions between the shoe and walkway are probably the most critical biomechanical parameters in slipping accidents. The ground reaction forces at the heel COP and opposing forces (Figure 2.1) determine whether a situation is potentially dangerous. When the ratio of horizontal and vertical force components (F_H/F_V) exceeds the resisting frictional shear force ($F\mu$) in the shoe–floor interface divided by the normal force (F_N) perpendicular to the floor surface, then a slip may commence. Note that, although not shown in Figure 2.1, the lateral force component also plays a role in slipping. The minimum friction requirement (μ_{min}) at equilibrium of forces is determined by the equation $\mu_{min} = F\mu/F_N = F_H/F_V$, where the former ratio represents the measured or "available" friction coefficient in the shoe–floor interface and the latter the reaction force ratio during gait (also see Section 2.5.1.1). The force ratio F_H/F_V has been termed the "required" friction coefficient on a nonslip surface (Redfern et al. 2001), when the foot is stationary with respect to floor, and "utilized" friction coefficient where motion may be present (Strandberg and Lanshammar 1981; Strandberg 1983a).

2.2.2.2 *Critical Gait Phases*

Gait is a repetitive event, with one cycle (one stride or two steps) used as the time period to describe the movement illustrated in Figure 2.2. The stride begins with the right foot in stance phase and proceeds from heel contact to midstance and finally toe-off. The left foot is in swing phase, but approaches heel contact when the right foot pushes the body forward. The horizontal (F_H) and vertical (F_V) ground reaction force components and their ratio (F_H/F_V) are subject to large variations during the stance as shown in Figure 2.2.

From the slipping point of view, two critical gait phases exist in level walking (Perkins 1978; Strandberg and Lanshammar 1981; Perkins and Wilson 1983; Strandberg 1983a). First, the early heel contact (peaks 3 and 4 in Figure 2.2) — when the rear part of the heel region is in contact with the ground, and second, the moment of toe-off (peaks 5 and 6) — when only sole forepart contact occurs. The first two main peaks of the ratio F_H/F_V in the forward direction of gait occur after 70–150 ms (10–20% into stance), depending on stance duration, before full body weight is transferred to the supporting foot. Peaks 1 and 2 are not considered hazardous because F_V is small at peak 1 and because peak 2 is directed backward.

The heel contact phase is considered more challenging for stability and more hazardous with regard to slipping than toe-off. This is because the forward momentum maintains the body weight on the supporting foot, with the potential for a forward slide to occur at peaks 3 and 4. A backward slip at toe-off is more easily prevented because most of the body weight is already transferred forward away from the slipping rear foot at peaks 5 or 6 to the leading foot.

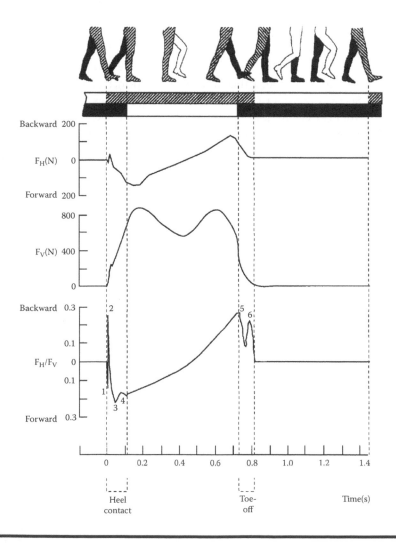

Figure 2.2 **Gait phases in normal level walking with typical horizontal (F_H) and vertical (F_V) ground reaction force components and their ratio, F_H/F_V for one step (right foot). (From Grönqvist, R., Roine, J., Järvinen, E., and Korhonen, E., 1989, *Ergonomics, 32,* 979–995. Adapted from Murray, 1967, *American Journal of Physical Medicine, 46,* 290–333 and Perkins, P.J., 1978, *Walkway surfaces: measurement of slip resistance,* ASTM STP 649 (Baltimore: American Society for Testing and Materials), pp. 71–87.)**

2.2.2.3 Anticipation and Adaptation to Slipperiness

A person may readily adapt to a slippery low-friction condition by adopting a protective gait strategy. This adaptation involves the combined effect of force and postural changes in the early stance, reducing the vertical acceleration and forward velocity of the body (Llewellyn and Nevola 1992). The body's COM is moved forward and closer to the BOS, step length is shortened and knee flexion increased, facilitating a more gentle foot contact with the ground. The ankle plantar flexion is greater than during a normal heel landing, thus reducing the contact angle of the shoe bottom with the floor. Thus, the shoe–floor contact area increases, while the shear forces decrease, resulting in a reduced utilized friction coefficient between shoe and floor, safer gait, and lower risk of injury. Other researchers (Marigold and Patla 2002; Cham and Redfern 2002) have verified that an individual's anticipation and prior experience of slipping can result in gait adjustments and modifications of the foot contact, so as to traverse a slippery surface safely. Cham and Redfern (2002) also reported that anticipating slippery flooring could produce significantly lower (16–33%) required friction coefficients than without anticipation.

2.3 Biomechanics of Slips, Trips, and Falls

This section describes what happens after the onset of a foot slide or a stumble during walking and during subsequent attempts by a person to bring the body back to equilibrium without falling.

2.3.1 Recovery of Balance and Risk of Injury

Recovery of balance and fall avoidance may be possible if a foot slide can be controlled and eventually stopped, or when a trip is followed by a successful compensatory stepping response, within a short enough time period (Grönqvist et al. 2001c, 2001d). Other possible recovery mechanisms for slips and trips may involve grasping responses, if any graspable fixtures are present. These may be combined with joint moment and stiffness control reactions, involving the ankle, knee, and hip joints in particular (Redfern et al. 2001). Marigold and Patla (2002) demonstrated that the reactive response to an unexpected first slip perturbation on rollers, comprised a rapid onset of a flexor synergy, a large arm elevation strategy and a modified swing limb trajectory; however, there may be an underlying and not yet fully understood risk associated with these responses. Movements made to restore balance and prevent falling have been found to create substantial muscle forces and unexpected loading of body parts (Lavender et al. 1988). Overexertion injuries to the spine and lower back, for example, may be caused by recovery attempts in response to slips and trips (Manning et al. 1984; Strandberg 1985; Stobbe and Plummer 1988).

2.3.2 Slip vs. Trip-Related Falls

The common factor in falls due to slipping and tripping is that a person's recovery attempts for controlling posture fail to restore balance before contact between body and the ground (or other object) occurs. Slipping in particular, but also tripping in some circumstances, may be related to the frictional characteristics of shoe soles, floor surfaces, and contaminants. Tripping and stumbling have more to do with uneven or discontinuous surfaces or foot contact with obstacles on the floor, however, resulting in unstable and erratic foot trajectories (Grönqvist et al. 2001c).

A trip may occur when the foot is suddenly interrupted during the swing phase (Redfern and Bloswick 1997). The swinging foot may fail to adequately clear the ground or it may encounter an obstacle or irregularity on the ground surface (Winter 1991). Grabiner et al. (1993), who studied the kinematics of recovery from an anteriorly directed stumble induced by an obstacle to healthy young males, found that the perturbation caused an increase in the maximum trunk flexion angle. This change was significantly associated with walking velocity and maximum hip and knee flexion angles, which increased due to the perturbation.

The most common protective movement outcome in response to an early swing perturbation, induced by an obstacle to healthy young men, according to Eng et al. (1994), is an elevating strategy (i.e., flexion of the swing limb). In response to a late swing perturbation, a rapid lowering of the swing limb to the ground and a shortening of the step length tends to occur (Eng et al. 1994). The late swing perturbation appears to pose a greater threat for a fall, because the body mass is already anterior to the stance foot. In this case, the elevating strategy is not possible due to the forward momentum of the body, precluding the stance leg from restoring the body equilibrium. The only option for recovery is to use the swing leg, if the hands have nothing to grasp.

2.3.3 Latencies of Muscle Responses to Slips, Trips and Stumbles

Corrective reactions to stumbling during gait have been studied using electromyograms of lower leg muscles to treadmill acceleration/deceleration or tibial nerve stimulation (Berger et al. 1984). For normal young adults, latency times for responses were 65–75 ms to treadmill acceleration/deceleration and approximately 90 ms to tibial nerve stimulation. In studies of movement strategies and muscular responses for recovery from a tripping perturbation, Eng et al. (1994) found that latencies of the reflex responses were 60–140 ms, suggesting that quick polysynaptic pathways are involved. Tang et al. (1998) observed that in healthy young adults, the latency times for reactive balance control to a slip perturbation (induced by a translational movement of a force plate) consisted of an early (latency time 90–140 ms) and high-magnitude activity of leg and thigh muscles. The bilateral anterior leg muscles as well as the anterior and posterior thigh muscles, with coordination between the two lower extremities, contributed to regaining balance within one gait cycle.

2.3.4 *Protective Movements in Falls*

When examining forward falls on an outstretched hand from low heights, Chiu and Robinovitch (1998) predicted that fall heights greater than 0.6 m carry a significant risk for wrist fractures to the distal radius (the most common type of fracture in the under-75 population). Robinovitch et al. (1996) and Hsiao and Robinovitch (1998) studied common protective movements that govern unexpected falls from standing height. They measured body segment movements when young subjects were standing on a mattress and attempted to prevent themselves from falling after the mattress was made to move abruptly. The subjects were more than twice as likely to fall after anterior translations of the feet (posterior fall) when compared with lateral or posterior translations (anterior falls).

Because a posterior fall would most likely follow a foot slide after heel landing, the study by Hsiao and Robinovitch (1998) may give an indication of possible mechanisms and protective movements for slip-related falls. Their results suggested that body segment movements during falls, instead of being random and unpredictable, involved a repeatable series of responses facilitating a safe landing. Posterior falls involved pelvic impact in more than 90% of Hsiao and Robinovitch's experiments, but only in 23% of the lateral falls and none of the anterior falls. In the falls that resulted in impact to the pelvis, a complex sequence of upper extremity movements allowed subjects to impact their wrist at nearly the same instant as the pelvis, suggesting a sharing of contact energy between the two body parts. Hsiao and Robinovitch (1999) predicted that a successful recovery from perturbation is governed by a coupling between step length, step execution time, and leg length.

2.4 Frictional Mechanisms in Walking

Frictional mechanisms in the context of walking have been discussed extensively by Tisserand (1985), Strandberg (1985), Proctor and Coleman (1988), Grönqvist (1995a, 1995b), and Leclercq et al. (1995a). Recently, Chang et al. (2001b) summarized the work in this field, including friction mechanisms at the shoe and floor interface on dry, liquid, and solid contaminated surfaces, and on icy surfaces. An overview of tribophysical mechanisms involved in slipping accidents is presented in a flowchart friction model (Figure 2.3).

The model takes into account the drainage capability of the shoe–floor contact surface (related to the squeeze-film process), draping of the shoe bottom about the asperities (microscopic irregularities) of the floor surface (related to hysteresis and lubrication), and finally the true contact between the interacting surfaces (related to traction and adhesion). The squeeze-film process, occurring between the shoe and the walking surface immediately after heel contact, is a critical phenomenon affecting balance and safety in gait when surfaces are subject to contamination. Thus, one of the key elements in slip and fall injury prevention is the drainage capability of walkways, floorings, and footwear solings.

In this context, the term "drainage" describes how effectively a shoe bottom is draped about the asperities of a floor surface, so that deformation and true contact may be obtained in the presence of a contaminant. The term "squeeze-film" (Figure 2.4 in this section) applies to the case of approaching surfaces, such as a shoe heel and floor, which attempt to displace a viscous fluid between them under the influence of a force (Moore 1972). The term "sinkage" (Figure 2.5) is often used to describe the descent of one of the bodies (in this case the shoe) when submerged in the fluid. Shoe–floor friction caused by "hysteresis" and "adhesion" is a manifestation of the same viscoelastic energy dissipation mechanism (Kummer 1966). Adhesional friction is caused by a dissipative stick–slip process at a molecular level, whereas hysteresis friction involves an irreversible and delayed response during a contact stress cycle. Hysteresis friction arises because when stress is removed in an elastic shoe soling during travel across a rigid surface, the soling does not recover completely to its original shape. This results in asymmetric deformation and pressure distribution during sliding at the contact interface. For a complete explanation of terms used in the flowchart friction model (Figure 2.3), readers are referred to Grönqvist (1995a, 1995b), Chang et al. (2001b), and Grönqvist et al. (2001c).

If drainage and draping after heel contact, involving a dynamic loading condition, is too slow due to hydrodynamic load support and elastohydrodynamic effects (as h in Figure 2.4 and Figure 2.5 approaches zero), the

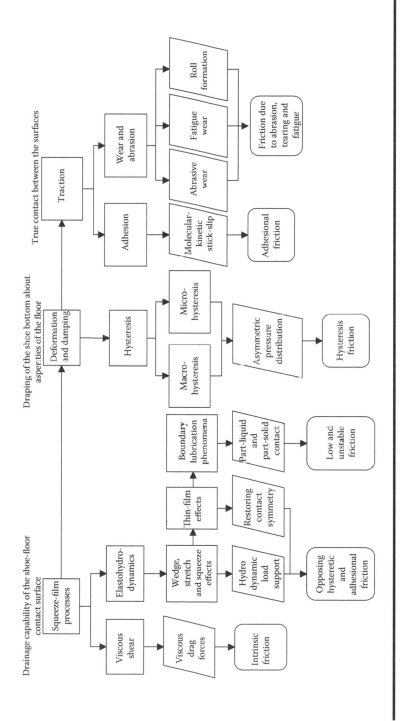

Figure 2.3   Flowchart friction model for slipping. (From Grönqvist, R., 1995a. *People and work, research reports 2* (Helsinki: Finnish Institute of Occupational Health), doctoral dissertation.)

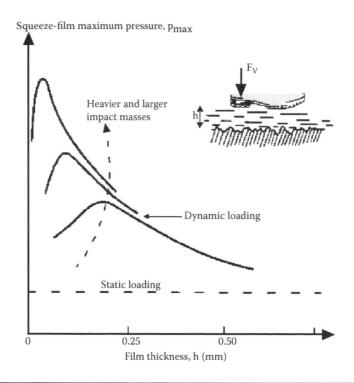

Figure 2.4   Squeeze-film maximum pressures (p_{max}) for impact and static loading. F_V = vertical component of ground reaction force and h = contaminant-film thickness. (Adapted from Moore, D.F., 1972. *International series of monographs on material science and technology*, Vol. 9 (Oxford: Pergamon Press).)

frictional forces in the shoe–floor contact surface will develop insufficiently to avoid slipping. In the opposite case when draping is quick enough, however, adequate frictional forces may develop due to deformation and damping (macro- and micro-hysteresis) and true molecular contact (adhesion and wear) between the interacting surfaces. Thus, a safer situation occurs.

Both the horizontal shear force (F_H) and vertical impact force (F_V) applied by the foot against the ground (Figure 2.1 and Figure 2.2) affect the risk of slipping as explained in Section 2.2. Smaller impact masses (and smaller F_V), in the presence of a contaminant-film (e.g., water or oil) tend to reduce the maximum hydrodynamic pressure (p_{max}) in the squeeze-film compared with larger impact masses (Moore 1972). The reduced p_{max}, which is also determined by the shoe–floor contact area and contaminant viscosity as film thickness (h) approaches zero, tend to reduce the load support enabling both hysteresis and adhesional friction to come into play in the shoe–floor contact surface. Thus, the result of more gentle heel contact is a higher friction coefficient, which leads to a reduced risk for

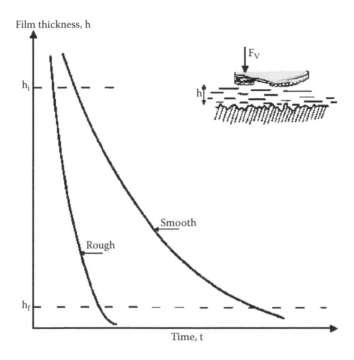

Figure 2.5 **Sinkage curves on rough and smooth surfaces.** F_V = **vertical component of ground reaction force,** t = **sinkage time,** h = **contaminant-film thickness** (h_i **is the initial and** h_f **the final film thickness). (Adapted from Moore, D.F., 1972.** *International series of monographs on material science and technology,* **Vol. 9 (Oxford: Pergamon Press).)**

slipping and falling. Note that the maximum contaminant-film pressure is independent of film thickness for the case of static loading (Figure 2.4), such as during quiet standing.

2.5 Basic Safety Criteria for Walking

2.5.1 Friction Requirements and Thresholds

2.5.1.1 Available Friction vs. Required or Utilized Friction

Measured frictional properties of footwear and walking surfaces (also termed "available" friction) can be determined using different slip measurement methodologies (discussed in Section 2.6). Task-, surface-, and human-related factors (e.g., walking, carrying, contamination, speed and cadence of gait, step length, anticipation, and adaptation) are known to affect balance, stability, and risk of slipping (Redfern et al. 2001; Grönqvist et al. 2001d). The usefulness of measured friction values in the assessment of slipperiness will depend, for each particular situation being evaluated, on the validity and reliability of the applied test method (Strandberg 1985; Chang et al. 2001b, 2001c). By comparing measured frictional properties of shoes and floor surfaces, i.e., available friction, with required or utilized friction requirements (defined in Section 2.2.2.1), it is possible to determine whether a situation in question is safe or dangerous (Redfern et al. 2001; Grönqvist et al. 2001c).

Safe thresholds for the friction coefficient in level walking have been found to vary between 0.15 and 0.30 (Perkins 1978; James 1980; Strandberg and Lanshammar 1981; Skiba et al. 1983; Strandberg 1983a; Strandberg et al. 1985; Redfern et al. 2001; Grönqvist et al. 2001c). Strandberg and Lanshammar (1981) measured the maximum utilized friction, peak F_H/F_V, approximately 0.1 seconds after heel contact. The peak value was on average 0.17 when there was no sliding between shoe and floor, 0.13 when the subject was unaware of the sliding motion or regained balance, and 0.07 when the slip resulted in a fall. Kinetic friction properties appeared to be more important than static, because in most walking experiments by Strandberg and Lanshammar (1981) the heel slid upon first contact, even on dry surfaces with no lubrication.

Whereas the significance of the required friction coefficient (maximum F_H/F_V) is that it indicates where in the gait cycle a slip is most likely to commence, the utilized friction coefficient has been used to determine the evolution of friction during the period of shoe–floor contact (Strandberg et al. 1985; Grönqvist 2001a). The latter criterion also makes possible prediction of whether it could be possible to retard or stop a slip that has already commenced. This is a key factor for successful recovery of balance after onset of a slip.

2.5.1.2 Difference between Static and Kinetic Friction

Using a simple biomechanical model (involving consideration of mechanical forces at the moment the foot strikes the ground), Tisserand (1985) suggested that slipping velocity (v) is a function of the difference between the static (F_s) and kinetic (F_k) friction forces:

$$v \approx \frac{1}{M}\left(F_s - F_k\right) \cdot t$$

where M is the mass of the body parts in motion and t is time.

Tisserand concluded that for a given coefficient of kinetic friction the seriousness of a fall would be directly proportional to the coefficient of static friction. He also presented experimental subjective evaluation data in support of his reasoning and went further in his analysis by assuming that this relationship would be valid even if no initial static phase exists as the heel strikes the ground. In conclusion, he stated that "preventing initial slipping of the foot requires a high or sufficient static friction force, while limiting slip velocity to avoid loss of balance requires a small difference between static and kinetic friction forces and that the latter is always required to prevent falling and injury."

2.5.2 Other Safety Criteria

2.5.2.1 Slip Velocity and Distance

Strandberg and Lanshammar (1981) estimated that the critical sliding velocity leading to falling after a heel slip was around 0.5 ms⁻¹, and that the required minimum kinetic friction coefficient was about 0.2 during normal level walking. If the preceding figures are accepted as critical for a hazardous slip and fall, then the boundary slip distance *s* between an avoidable and unavoidable fall is approximately 60 mm (Grönqvist et al. 2001d) because:

$$s = \frac{v^2}{2g\mu}$$

The preceding equation, where *g* is the acceleration of gravity (1 *g* = 9.81 ms⁻²), *v* is velocity of sliding, and μ is the coefficient of friction, is derived from the work done by the frictional force. This equation, governing the distance required to bring a moving object to a stop by friction, is based on the assumption of a constant initial sliding velocity, which may not hold in practice. Nevertheless, it indicates that if the coefficient of friction increases for example to 0.4, then the boundary slip distance would be reduced to 30 mm. This distance is likely to be perceived as a slip by only 50% of subjects walking over a floor surface (Leamon and Li 1990). Thus, a safer situation from the point of view of balance recovery will exist. If the critical values (*s* = 0.2 m and *v* = 1.1 ms⁻¹) for slip distance and velocity recently suggested by Brady et al. (2000) are inserted in the equation, then the minimum coefficient of friction in the interface would need to be around 0.3 for fall recovery.

Winter (1991) and Lockhart (1997) reported a greater horizontal heel contact velocity for older vs. younger subjects on dry floor surfaces, even though the walking velocity of the older subjects was slower. On a slippery floor surface, however, a higher heel contact velocity coupled with a slower transition of whole body COM velocity of older subjects significantly affected sliding heel velocity and dynamic friction utilization. The result was longer heel slip distances and increased falling frequencies for older compared with younger individuals (Lockhart et al. 2000a, 2000b). Recent findings of Lockhart et al. (2002) indicate that sensory changes in the elderly may increase the likelihood of slips and falls compared with their younger counterparts, due to incorrect perceptions of floor slipperiness and poorer gait adaptation when slippery conditions are encountered.

2.5.2.2 Trunk Acceleration

As slipping may involve transient movements resulting from a person's efforts to regain balance, some researchers have focused on understanding the role of such unexpected movements for maintaining balance in walking (Hirvonen et al. 1994, 1996). Unexpected movements for regaining balance can be discriminated and measured in terms of antero-posterior and medio-lateral trunk accelerations during slip and trip incidents, with normal movement signals attenuated using band-pass filtering. Hirvonen et al. (1994) monitored unexpected trunk movements during slip incidents of twenty male volunteers, walking at two speeds (normal and race walking) along a horizontal track. Peak acceleration levels of the trunk increased significantly in slipping incidents compared with normal or race walking without slipping, both in the antero-posterior and medio-lateral directions. Peak accelerations varied from 0.5–4.5 g during slipping (most experiments resulted in either controlled or vigorous slips), while the accelerations were less than 0.5 g during walking without slipping.

2.6 Methods for Slipperiness Measurement

To be able to measure the exposure to slip hazards and to assess the risk of injury, it is necessary to use a valid and reliable methodology. Numerous techniques have been proposed, with approaches that are either human-centred or apparatus-based. Apparatus-based methods are more practical for routine purposes, such as for product testing in the laboratory and for monitoring risk exposures at worksites. Human-centred approaches, on the other hand, are inherently valid and include human factors aspects as one part of the risk assessment. Importantly, they allow a combination of biomechanical measurement along with subjective observation of performance. Human-centred field applications are rare and time-consuming, however, compared with the use of portable slip testing devices.

2.6.1 Human-Centred Approaches

Human-centred approaches for slipperiness measurement may be psycho-physical in nature. A perceived magnitude of "slipperiness" can be quantified on a psychophysical scale using "foot movement" or "postural instability" as the physical stimulus (Grönqvist et al. 2001d). The stimulus can be measured subjectively using opinions and preferences but also objective measurements may be applied, for instance, by video recording or high-speed imaging of gait. Such approaches may also include acquisition of ground reaction force data (e.g., Strandberg 1985, Grönqvist et al. 1993).

A number of subjective measures of slipperiness (e.g., paired comparisons) have been applied to assess fall risk. Human subjects are capable of differentiating the slipperiness of walkway surfaces (Yoshioka et al. 1978, 1979; Swensen et al. 1992; Myung et al. 1993; Chiou et al. 1996) and footwear solings (Strandberg et al. 1985; Tisserand 1985; Nagata 1989; Grönqvist et al. 1993) in dry, wet, or contaminated conditions. Leamon and Son (1989) and Myung et al. (1992) suggested that measuring micro-slip length or slip distance during slipping incidents might be a better means of estimating slipperiness than many of the apparatus-based friction measurement techniques. Recently, Chiou et al. (2000) reported findings of workers' perceived sense of slipping during task performance while standing (e.g., a lateral reach task) and related their sensory slipperiness scale to subjects' postural sway and instability. They found that workers who were cautious in assessing surface slipperiness had less postural instability during task performance.

Skiba et al. (1986), Jung and Schenk (1989, 1990), and Jung and Rütten (1992) evaluated walking test methods used in the laboratory for measuring the slipperiness of floor coverings and safety footwear on an inclined plane in dry, wet, and oily surface conditions. The inclination angle at the point when walking down the ramp becomes unsafe can give an estimate for slip resistance by transforming it geometrically to a friction value. These articles also discussed the validity and reliability of such tests, use of standard reference materials and separation characteristics for choosing a limited number of test subjects for standardised slipperiness measurements.

Subjective measurement approaches have also been utilized to assess footwear friction on icy surfaces (Bruce et al. 1986; Manning et al. 1991). The test rig by Bruce et al. (1986) consisted of a tubular metal frame with four legs, fitted with large castor action wheels. A test subject, standing on both feet on an icy surface, was dragged across the substrate at a low sliding speed on an ice skating rink and the horizontal (frictional) force in the shoe–ice interface was measured using a load cell (spring balance). The frame of the test rig prevented the subject from falling. Manning et al. (1991) applied another technique to measure the slipperiness of a shoe–ice as well as normal shoe–floor interface. A subject walked backward on the surface to be assessed, while pulling against a spring. The subject was supported by a fall-arrest harness and protected by two handles suspended from a pulley, which moved freely on an overhead rail. The load cell was positioned between the harness belt of the subject and a rigid base (e.g., a wall), so that an indirect measure of shoe–floor friction was obtained. Manning and Jones (1993) subsequently modified this walking traction rig for mobile field use; however, Scheil and Windhövel (1994) criticised the validity and precision of this method because inertial forces in walking may significantly increase the load cell reading compared with the actual frictional force in the shoe–floor interface.

2.6.2 Apparatus-Based Approaches

In a survey of testing methods, Strandberg (1983b) found more than 70 different friction testers in existence at that time. Many of these laboratory and field based devices are still in use (Grönqvist 1995a, 1999). Current techniques typically measure either static friction properties of surfaces, limited to the start of a foot slide, or steady state kinetic friction properties, relevant to a prolonged slip condition (Grönqvist 1997; Grönqvist et al. 1999). Some devices comprise features, however, allowing the measurement of transitional kinetic friction properties of the shoe–floor interface by simulating the heel contact phase in gait (Grönqvist et al. 2001b).

A recent survey of friction measurement devices by Chang et al. (2001c) compared a number of laboratory and field based test methods in use at present, under various test conditions, using biomechanical observations. Device validity, repeatability, reproducibility, and usability were examined and compared against the published literature. The validity and reliability of most devices could be improved by better control of test parameters (i.e., buildup rate and magnitude of normal force, contact area and time, and sliding velocity), which should then better reflect the biomechanics of human slipping conditions.

2.6.3 Kinetic vs. Transitional and Static Friction

Due to the dynamics of the heel contact phase, slip resistive frictional properties between shoes and floors need to be considered during transition from static to dynamic conditions as well as during steady state dynamic conditions (Tisserand 1985; Grönqvist 1995a,b). The severity of a slip and fall incident depends on how quickly and to what extent slip resistive frictional forces build up during a short transition period (< 250 ms) after heel contact (Grönqvist et al. 2001a). A longer transition period (>250 ms) is required for measuring steady state kinetic friction properties of the shoe–floor interface in the presence of viscous contaminants (Grönqvist 1997). Because the events in slips that result in falling tend to evolve quickly, in less than half a second (Grönqvist et al. 2001c), it appears reasonable that steady-state kinetic friction measurements ought not be relied upon solely without considering transitional and static friction (Section 2.5.1.2). Corrective movements for balance recovery must immediately follow a slip perturbation if a fall and subsequent injury are to be avoided. Consequently, a need exists for the measurement and consideration of contact-time-related frictional variation between shoes and floors to allow accurate risk assessment for slipping.

2.7 Implications for Falling Accidents and Guidance for Prevention

2.7.1 Surface Roughness and Porosity Effects

In addition to direct measurement of slipperiness, using the approaches described previously, measurement of various surface roughness characteristics can also give a valuable indication of the likely performance of different footwear or flooring surfaces with regard to slipping. Surface roughness has a substantial effect on friction and slip resistance, particularly when liquid or solid contaminants are present in the shoe–floor interface. The drainage capability of the shoe soling through the contaminant-film is dependent on the macro- and, in particular, micro-roughness of the interacting shoe and floor surfaces (Chang et al. 2001a). The displacement of contaminants in the shoe–floor interface affects how effectively friction can be built up during the approach of the surfaces. The time of the approach can be quantified by the rate of sinkage (Moore 1972). Figure 2.5 gives an example of sinkage rates for a rough and a smooth surface. A quicker rate of sinkage facilitates higher traction and safer human locomotion in the presence of contaminants (see Section 2.4). Smooth surfaces are in general more hazardous than rougher surfaces, but the type of contaminant (liquid vs. solid) and its viscosity (water vs. oil) greatly affects the slipperiness of the shoe–floor interface.

A thorough analysis of the role of surface topography changes and surface roughness in slipperiness measurement can be found in recent articles by Kim and Smith (2000) and Chang et al. (2001a). Kim and Smith observed surface topography changes in shoe solings and floor surfaces

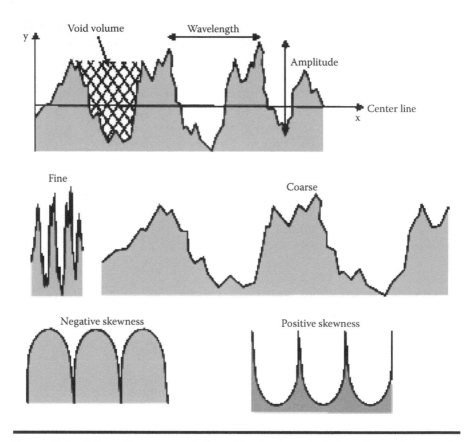

Figure 2.6   Schematic representation of basic characteristics of a surface roughness profile: roughness amplitude, wavelength, and skewness.

during relative sliding in dry conditions and showed how these geometric characteristics are continuously modified by wear processes affecting slip resistance performance.

A schematic representation of some basic characteristics of a surface roughness profile is given in Figure 2.6. The surface amplitude parameters relate to heights of peaks and depths of valleys in a surface profile (e.g., maximum of peak to valley height, R_y). Amplitude parameters R_a, R_{tm}, and R_z (DIN) are defined in Section 2.7.1.1. The spatial wavelength parameters (e.g., mean spacing of adjacent local peaks, S, and root mean square of spatial wavelength, λ_q) are indicators of the density of adjacent profile heights. The slope parameters (e.g., root mean square of surface slope, Δ_q) are related to the contact angle between surface asperities, when a shoe sole slides over a floor surface, where tangential shear forces are present (Kim and Smith 2000). Two important surface parameters — skewness, R_{sk}, and kurtosis, R_{ku} — are measures of the shape and spikiness of surface profiles (Chang et al. 2001a). Symmetric profiles with equal numbers of peaks and valleys have zero skewness. A surface profile with negative skewness has peaks with broad plateaux and narrow valleys (small void volume), whereas the counter-profile shape represents positive skewness, which is characterized by wide valleys (large void volume) and narrow plateaux.

Although adhesional friction dominates on dry surfaces, friction due to hysteresis is the main component on wet and contaminated surfaces, as described previously. Hysteresis friction, which is proportional to the interface contact pressure, is enhanced on finer surfaces (small values of λ_q) with larger slope (Δ_q) due to an increased deformation frequency and hysteresis loss in the shoe–floor contact surface. The deformation frequency is defined as sliding velocity divided by distance between surface asperities (Moore 1972). Viscous lubricants, such as oil, require a coarser floor surface texture and a greater void volume than wet conditions, for example, for slipperiness to be reduced.

2.7.1.1 Walkway Surfaces

Research findings on the influences of walkway surface roughness on slipperiness were first reported by Jung and Riediger (1982), Grönqvist et al. (1988b, 1990), Harris and Shaw (1988), and Wieder (1988). Jung (1991) summarized the required minimum values for flooring roughness parameters R_a (arithmetic average of surface heights, also known as centre line average), R_z (DIN) and R_{tm} (average of peak to valley height) in several lubricant conditions by transforming the parameter values to R_z (DIN) values. Note that these parameters only describe the roughness amplitude characteristics of a surface profile (Figure 2.6), and that different friction measurement techniques were applied to measure slip resistance. Nevertheless, the suggested minimum requirements expressed using R_z (DIN) were as follows: 7.6–8.8 µm for damp surfaces, around 34 µm for wet surfaces, 31–39 µm for surfaces contaminated with concentrated glycerol, and 34–140 µm for oily surfaces. Examples of typical roughness characteristics for some common floor surfaces are presented in Table 2.1.

The preceding figures must be taken as only crude guidelines for selecting floor surface roughness profiles for different environments. Other factors need to be considered too, including porosity (permeability) and the effects of cleaning. Grönqvist et al. (1992), through measuring the correlation between floor surface roughness (R_a) and slip resistance of a number of ceramic tiles and polymeric resin floorings, observed that in addition to surface roughness, porosity of surfaces also appeared to enhance slip resistance. Porous floorings (concrete surfaces and unglazed

Table 2.1   Examples of Typical Roughness Characteristics for Some Common Floor Surfaces

	Surface Roughness Parameter			
Floor Surface	*Amplitude (μm)[a]* R_a	*Wavelength (μm)[b]* λ_q	*Slope (degrees)[c]* Δ_q	*Skewness[d]* R_{sk}
Stainless steel (smooth)	0.15	120	0.55	+0.90
Glazed ceramic tile (smooth)	0.75	600	0.55	−0.01
Semiglazed ceramic tile (rough)	17.0	670	12.0	+1.30
Unglazed ceramic tile (semirough)	2.90	150	9.20	−0.50
Vinyl tile (smooth)	1.50	—	—	—
Antislip vinyl with silicon carbide particles	7.50	—	—	—
Concrete with acrylic resin coating	6.00	—	—	—
Epoxy resin with quartz particles (semirough)	5.00	—	—	—
Acrylic resin with quartz particles (rough)	14.0	—	—	—
Abrasive safety floor tape (rough)	22.0	—	—	—

— = missing parameter value.

[a] Larger amplitude indicative of better slip resistance in wet or oily conditions.

[b] Smaller wavelength indicative of better slip resistance in wet or oily conditions.

[c] Larger slope indicative of better slip resistance in wet or oily conditions.

[d] Symmetric profiles with equal numbers of peaks and valleys have zero skewness (cf. Figure 2.6); positive skewness provides larger void volume for better contaminant displacement, however, negative skewness provides larger apparent contact area.

ceramic tiles) gave the highest friction readings in the presence of two different contaminants (soap solution and concentrated glycerol), despite their significantly smaller roughness amplitudes in comparison to several other measured surfaces. Leclercq et al. (1997) confirmed that the permeability of a floor surface is important in terms of kinetic friction and slip resistance. They found that slip resistance in a mechanics workshop was higher on permeable surfaces (concrete) than on impermeable surfaces (resins) covered with oil. Moreover, the slip resistance decreased on permeable concrete surfaces immediately after cleaning but increased for impermeable resin floors. To explain this, Leclercq et al. (1997) suggested that the impermeable surfaces were more easily degreased during the cleaning process than the permeable surfaces, which became soaked with the contaminant.

Lloyd and Stevenson (1992) presented a comprehensive floor surface "roughness index" (RI) based on a model combining the surface profile's amplitude (R_q), wavelength (λ_q), and skewness (R_{sk}). They obtained a statistically significant combined linear correlation coefficient (0.983) between the RI and measured coefficients of kinetic friction among a number of shoe–floor surfaces, subject to oil contamination.

Chang (1998, 1999, 2001) studied extensively the effect of floor surface roughness on friction and slip resistance. Chang (1998) obtained the highest correlation between roughness and friction with the following surface parameters: average of maximum heights above the centre line (R_{pm}) and arithmetic mean of surface slope (Δ_a). The former represents the averaged void volume among surface asperities available to contain contaminant, whereas the latter is related to the rate of asperity deformation during sliding contact. In wet conditions, sharper and higher peaks with an optimal high peak density increase the friction coefficient according to Chang (1999). Chang (2001) concluded that a good indicator of friction will most likely comprise the surface parameters representing surface slope, peak-to-valley distance, and surface void volume.

Chang and Matz (2000) and Chang (2000) investigated the effect of selection of filter type and bandwidth on various outcome parameters in surface roughness measurement with a stylus profilometer. They recommended using filter type 2CR PC and measuring parameters Δ_a, R_{pm}, and R_k (kernel roughness depth) for the purpose of comparing floor surface roughness with the measurement of friction. Measurement of surface roughness usually requires removing the waviness portion from the measured profiles, however, which may limit the interpretation of the results for prevention purposes. Considering this, Chang et al. (2004) suggested that in contrast to the current common practice of measuring surface roughness, surface waviness should be measured when a short cutoff length (0.8 or 2.5 mm) is used or when the viscosity of the contaminant is high.

2.7.1.2 Footwear Solings

Roughness of footwear soling materials also affects friction and slip resistance. Manning et al. (1983) tested the slip resistance of footwear throughout its useful life. Footwear solings became polished during walking on oily and wet floor surfaces, accompanied, in general, by a changing pattern of slip resistance. Manning et al. observed that the polishing effect was less apparent on polyurethane soles than with nitrile rubber specimens. Manning et al. (1990) measured shoe soling slip resistance on water-lubricated and oily floor surfaces (e.g., stainless steel, smooth vinyl, and rough plastic). They reported that abrasion of rubber soling materials with different grades of abrasive paper (using an orbital sander) gradually raised the value of R_{tm}-roughness from 3.6–6.5 µm to 12.3–25.4 µm, with slip resistance improved. Both roughness and slip resistance decreased during subsequent polishing of the soling materials.

Rowland et al. (1996) contributed to work done in this field by comparing scanning electron microscopic images with surface roughness measurements. They concluded that the cellular surface structure of microcellular rubber and microcellular polyurethane differed considerably from that of dense soling materials (e.g., nitrile rubber), which had a relatively smooth surface.

Manning et al. (1998) conducted an experiment with five pairs of shoes soled with the same rubber compound. Four of the pairs were abraded by different grades of grit to produce a range of roughness values (R_{tm} 7–19 µm). One pair had a newly moulded surface polished to 4.4 µm. Friction coefficients of the shoe pairs were measured on two atypically smooth float glass surfaces (R_{tm} 0.1 µm) and three relatively smooth floor surfaces (R_{tm} 0.7–6.9 µm) under water-lubricated conditions. The measured friction coefficients varied from 0.07–0.34 and the results demonstrated that soling roughness was a major factor in determining the wet friction coefficient, although the surface roughness of the floors also had a significant effect on friction. Manning and Jones (2001) summarized their research on the effects of surface roughness over a period of 18 years and concluded that a microcellular polyurethane compound AP66033 is the most slip-resistant safety footwear material on oily and wet surfaces.

Grönqvist et al. (1988a) and Grönqvist (1995b) compared shoe-soling roughness with the coefficient of kinetic friction on two contaminated (water with detergent and concentrated glycerol) floor surfaces (smooth steel and rough plastic). The roughness was measured in terms of R_a-values for three different types of shoe heels made of compact nitrile rubber (NBR, density 1.2 g cm^{-3}), compact styrene rubber (SBR, density 1.1 g cm^{-3}), and microcellular polyurethane (polyurethane [PU], density 0.6 g cm^{-3}). The results showed that in each material category, the highest friction readings were obtained with an optimum combination of tread-pattern depth (2.6–3.1 mm) and surface roughness (R_a 5–9 µm) of the heel. However, the only type of footwear that obtained an overall safe level of the friction coefficient ($\mu > 0.2$) in all test conditions had heels made of microcellular PU (tread-pattern depth, 3.1 mm, and surface roughness, R_a 7 µm).

Abeysekera and Gao (2001) did not find any significant correlation between the measured friction coefficient and shoe soling roughness (R_a 4.3–24.1 μm) on pure and warm ice. Their finding indicates that factors other than soling roughness determine the friction coefficient on warm ice when the surface temperature is close to 0°C (footwear interaction with ice is discussed further in Section 2.7.3).

2.7.2 Shoe Soling Material and Tread Wear Effects

Strandberg (1985) and Strandberg et al. (1985) examined the effects of a number of physical soling properties on utilized friction and frequency of falling. This involved trained subjects walking over different contaminated floor surfaces in a triangular closed path, as fast as possible, without slipping and falling into a safety harness. The results indicated that a number of properties (e.g., soling material, tread-pattern depth, texture, hardness, and flexibility) affect drainage, draping and true contact between shoe soling and floor in the presence of liquid and solid contaminants. Shoe solings without tread-pattern or with closed cavities should be avoided, according to Strandberg et al. (1985), due to insufficient drainage. Coarse textures in the sole may cause a small draping area in the shoe–floor contact and thus can contribute to a poor slip resistance.

Chiou et al. (1996) investigated the relationship between shoe usage and shoe wear. They determined how much tread is still available after some shoe usage (by taking a shoe print) and found that increasing shoe hardness increased the available tread-pattern. They also emphasized the need to consider the effect of available tread-pattern for producing acceptable friction readings, so that slip potential could be minimized. Chiou et al. (1996) concluded that further research is needed to develop guidelines for when work shoes should be replaced to reduce slips and falls.

Material wear may have a considerable effect on the frictional properties of shoe solings. Frequently, heels and soles are more slippery in their unworn stage than after a short period of usage. Effects of soling wear on friction have been reported during and after wear trials by volunteers for a period of up to two years (Manning et al. 1985), or when test subjects walked several kilometres on an asphalt path (Leclercq et al. 1995b). Manning et al. (1985) found that microcellular PU soles always produced higher friction coefficients on oil compared with nitrile rubber, and that the PU soles typically showed an increase in friction during the first weeks of wear. Leclercq et al. (1995b) compared soles made of expanded polyurethane (microcellular PU) with soles made of compact PU elastomers and found that the relative increase in slip resistance due to product wear was higher for the expanded (33–100%) than for the compact (0–54%) PU materials on oily surfaces.

Grönqvist et al. (1988a) and Grönqvist (1995b) adopted an alternative approach and collected footwear specimens worn by shipyard workers who had exposed their shoe solings to normal wear at work during a period of 2–8 months. These worn shoe solings were classified as "good," "satisfactory," "poor," or "worn out" based on their appearance and their slip resistance was measured in terms of a kinetic friction coefficient. A comparison was made with similar "new" (never worn) shoe solings (Grönqvist 1995b). The average change in slipperiness for the worn (four categories) vs. the new shoe solings was quantified on two floor surfaces (smooth steel and rough plastic) using two contaminants (water with detergent and concentrated glycerol). The results were as follows: 66% increase in slip resistance (μ_k 0.216 vs. 0.130) for low-density microcellular PU, 27% increase (0.143 vs. 0.113) for high-density styrene rubber (SBR), and 7% decrease (0.098 vs. 0.105) for high-density nitrile rubber (NBR). Only the soling material made of microcellular PU maintained the improved friction level throughout the total wear-period of 8 months. The study recommended that even relatively good performing footwear with microcellular PU solings should be discarded and replaced before the tread-pattern is worn-out.

These results suggest that it may be possible to increase walking safety by "running-in" or slightly abrading the heel and sole surfaces of new shoes, before putting them on the market or before starting to wear them (e.g., Grönqvist et al. 1988a, Grönqvist 1995b). The best way of executing this run-in procedure needs further research, however, although the work of Manning et al. (1990) and Manning and Jones (1994, 2001) provides a useful indication. They suggested the use of a belt-sanding machine coated with P100 grit abrasive paper. Low-density soling materials such as micro-cellular PU, which are inherently rough, are more sensitive to traction improvement by abrasion than high-density materials, such as SBR and NBR solings, which gradually become polished after the abrasion, return-ing to a more slippery state.

2.7.3 Shoe Soling Hardness and Ice

When walking on ice, the thickness of any water layer and properties of the ice, such as its temperature, structure, and hardness, appear to be more important factors determining slipperiness, instead of the nature of footwear soling (Grönqvist 1995a; Chang et al. 2001b). Nevertheless, low hysteresis and low hardness, which are often interrelated, seem to be desirable properties of rubber like materials to improve friction in icy conditions (Ahagon et al. 1988).

The hardness of shoe soling is known to affect friction and slip resistance on cold ice (Bruce et al. 1986; Grönqvist and Hirvonen 1995).

Statistically significant relationships between friction coefficients in the range 0.31 decreasing to 0.03 and increasing soling hardness in the range 42–94 Shore A were found at low ice temperatures (−9 or −10°C). Bruce et al. (1986) suggested that the hardness of the footpads of a polar bear's feet (20–30 Shore A) might represent the optimum hardness for shoe heels and soles on ice.

Warm ice (close to 0°C) differs from colder ice with regard to the influence of soling hardness, however. Grönqvist and Hirvonen (1995) found that on warm ice, which was significantly more slippery than cold ice, only hard soling materials with sharp cleats improved the friction coefficient to some extent, because the cleats scratched the ice surface. The effect of softer heel and sole materials on the friction coefficient was negligible on warm ice according to Grönqvist and Hirvonen. Recently, Abeysekera and Gao (2001) confirmed this, finding no significant correlation between the measured friction coefficient and shoe soling hardness for pure ice, having a surface temperature near to 0°C.

For other floor surfaces, the relationship between soling hardness and slip resistance is not clear and has been little studied. Manning et al. (1991), reporting soling hardness values and measured friction coefficients for a number of footwear types on five floor surfaces and on ice, did not find relationships between the two variables. Theoretically, one might anticipate that soling hardness plays a more important role on dry floor surfaces (due to higher adhesional friction for softer materials) than on wet or contaminated floors because friction due to deformation and hysteresis would then be the main component, as discussed previously in this chapter.

2.7.4 Antiskid Devices for Ice and Snow

Methods to describe functional problems in walking on different slippery surfaces during winter conditions have been developed by Gard and Lundborg (2000), with rating scales for evaluating walking safety and balance, and observation scales for monitoring posture and movements during walking. These methods have been used to investigate functional problems when wearing different antiskid devices (attached to shoes) for slip and fall protection. Gard and Lundborg (2001) performed practical tests of 25 different antiskid devices on the Swedish market, on different icy surfaces with gravel, sand, salt, or snow on ice, as well as with pure ice. They distinguished four principal designs for antiskid devices:

1. Fixed heel devices
2. Forefoot devices
3. Whole-foot devices
4. Heel devices

The tested devices were evaluated according to each subject's perception of walking safety, walking balance and likelihood for self-use. Each subject's posture and movements during walking were also analysed by an expert physiotherapist. One of the tested antiskid devices (a fixed heel device) was judged good regarding walking safety and balance and was also selected by the subjects for their own preferred use (Gard and Lundborg 2001).

2.8 Conclusions

The most effective way to eliminate or reduce slipping or tripping hazards is to design safe environments that take account of the limits of human capacity. Measures that improve tidiness and orderliness are effective for preventing slip- and trip-related falls, because these often happen on unsatisfactorily maintained, wet or contaminated walkways or when obstacles or poorly fitting carpeting interrupt normal gait. In situations where risks remain present and walkers are aware of these, they may be able to adapt their gait to the circumstances, reducing risk of injury. This might involve changes that reduce the likelihood of slipping or tripping, improve the recovery from a slip or trip without falling, or enable specific protective movements to limit the consequences of a fall.

It has been explained that friction, wear, and surface roughness, along with shoe soling and flooring material properties, are the most important technical aspects affecting falls in slippery environments. In this respect, field monitoring of walkway slipperiness has an important role to play in fall-injury prevention. Either human-centred or apparatus-based approaches for slipperiness measurement may be used to evaluate on-site slipping risks; however, measurement requires trained operators, using recognised test instruments and proven methodologies.

Previous research in the field of slips, trips, and falls on the level has been reviewed, including kinematics and ground reaction forces in normal gait, postural balance, and fall biomechanics. Prevention needs a thorough knowledge of the underlying mechanisms and contributory factors for falls. Multidisciplinary research effort across the fields of biomechanics and motor control, tribophysics and materials science, ergonomics, cognitive psychology, and injury epidemiology, among others, is required to further improve understanding of the causes of falls, their consequences, and approaches to prevention.

Acknowledgments

The author is grateful to Ms. Milja Ahola and Ms. Armi Muhonen for their assistance in designing and drawing the artwork for the illustrations. The author expresses his sincere thanks to Mr. Jouko Rytkönen for helping with conversion of the electronic files.

References

Abeysekera, J. and Gao, C., 2001, The identification of factors in the systematic evaluation of slip prevention on icy surfaces. *International Journal of Industrial Ergonomics*, 28, 303–313.

Ahagon, A., Kobayashi, T., and Misawa, M., 1988, Friction on ice. *Rubber Chemistry and Technology*, 61, 14–35.

Berger, W., Dietz, V., and Quintern, J., 1984, Corrective reactions to stumbling in man: neuronal co-ordination of bilateral leg muscle activity during gait. *Journal of Physiology*, 357, 109–125.

Brady, R.A., Pavol, M.J., Owings, T.M., and Grabiner, M.D., 2000, Foot displacement but not velocity predicts the outcome of a slip induced in young subjects while walking. *Journal of Biomechanics*, 33, 803–808.

Bruce, M., Jones, C., and Manning, D.P., 1986, Slip-resistance on icy surfaces of shoes, crampons and chains — a new machine. *Journal of Occupational Accidents*, 7, 273–283.

Cappozzo, A., 1991, The mechanics of human walking. In A.E. Patla (Ed.) *Adaptability of human gait: implications for the control of locomotion* (Amsterdam: Elsevier/North-Holland), 55–97.

Carlsöö, S., 1962, A method for studying walking on different surfaces. *Ergonomics*, 5, 271–274.

Cham, R. and Redfern, M.S., 2002, Changes in gait when anticipating slippery floors. *Gait and Posture*, 15, 159–171.

Chang, W.R., 1998, The effect of surface roughness on dynamic friction between neolite and quarry tile. *Safety Science*, 29, 89–105.

Chang, W.R., 1999, The effect of surface roughness on the measurements of slip resistance. *International Journal of Industrial Ergonomics*, 24, 299–313.

Chang, W.R., 2000, The effect of filtering processes on surface roughness parameters and their correlation with the measured friction. Part II: porcelain tiles. *Safety Science*, 36, 35–47.

Chang, W.R., 2001, The effect of surface roughness and contaminant on the dynamic friction of porcelain tile. *Applied Ergonomics*, 32, 173–184.

Chang, W.R., Grönqvist, R., Hirvonen, M., and Matz, S., 2004, The effect of surface waviness on friction between Neolite and quarry tiles. *Ergonomics*, 47, 890–906.

Chang, W.R., Grönqvist, R., Leclercq, S., Brungraber, R.J., Mattke, U., Strandberg, L., Thorpe, S., Myung, R., and Makkonen, L., 2001c, The role of friction in the measurement of slipperiness. Part 2: survey of friction measurement devices. *Ergonomics*, 44, 1233–1261.

Chang, W.R., Grönqvist, R., Leclercq, S., Myung, R., Makkonen, L., Strandberg, L., Brungraber, R.J., Mattke, U., and Thorpe, S.C., 2001b, The role of friction in the measurement of slipperiness. Part 1: friction mechanisms and definitions of test conditions. *Ergonomics*, 44, 1217–1232.

Chang, W.R., Kim, I.-J., Manning, D.P., and Bunterngchit, Y., 2001a, The role of surface roughness in the measurement of slipperiness. *Ergonomics*, 44, 1200–1216.

Chang, W.R. and Matz, S., 2000, The effect of filtering processes on surface roughness parameters and their correlation with the measured friction. Part I: quarry tiles. *Safety Science*, 36, 19–33.

Chiou, S., Bhattacharya, A., and Succop, P.A., 1996, Effect of workers' shoe wear on objective and subjective assessment of slipperiness. *American Industrial Hygiene Association Journal*, 57, 825–831.

Chiou, S., Bhattacharya, A., and Succop, P.A., 2000, Evaluation of workers' perceived sense of slip and effect of prior knowledge of slipperiness during task performance on slippery surfaces. *American Industrial Hygiene Association Journal*, 61, 492–500.

Chiu, J. and Robinovitch, N., 1998, Prediction of upper extremity impact forces during falls on the outstretched hand. *Journal of Biomechanics*, 31, 1169–1176.

Eng, J.J., Winter, D.D., and Patla, A.E., 1994, Strategies for recovery from a trip in early and late swing during human walking. *Experimental Brain Research*, 102, 339–349.

Gard, G. and Lundborg, G., 2000, Pedestrians on slippery surfaces during winter — methods to describe the problems and tests of anti-skid devices. *Accident Analysis & Prevention*, 32, 455–460.

Gard, G. and Lundborg, G., 2001, Test of anti-skid devices on five different slippery surfaces. *Accident Analysis & Prevention*, 33, 1–8.

Grabiner, M.D., Koh, T.J., Lundin, T.M., and Jahnigen, D.W., 1993, Kinematics of recovery from a stumble. *Journal of Gerontology*, 48, M97–M102.

Grönqvist, R., 1995a, A dynamic method for assessing pedestrian slip resistance. In *People and work, research reports 2* (Helsinki: Finnish Institute of Occupational Health), doctoral dissertation.

Grönqvist, R., 1995b, Mechanisms of friction and assessment of slip resistance of new and used footwear soles on contaminated floors. *Ergonomics*, 28, 224–241.

Grönqvist, R., 1997, On transitional friction measurement and pedestrian slip resistance. In P. Seppälä, T. Luopajärvi, C.-H. Nygård, and M. Mattila (Eds.) *Proceedings of the 13th Triennial Congress of the International Ergonomics Association, from Experience to Innovation*, Vol. 3 (Helsinki: Finnish Institute of Occupational Health), pp. 383–385.

Grönqvist, R., 1999, Slips and falls. In S. Kumar (Ed.) *Biomechanics in ergonomics*, Chapter 19 (London: Taylor & Francis), pp. 351–375.

Grönqvist, R., Abeysekera, J., Gard, G., Hsiang, S.M., Leamon, T.B., Newman, D.J., Gielo-Perczak, K., Lockhart, T.E., and Pai, C. Y.-C., 2001d, Human-centred approaches in slipperiness measurement. *Ergonomics*, 44, 1167–1199.

Grönqvist, R. and Hirvonen, M., 1995, Slipperiness of footwear and mechanisms of walking friction on icy surfaces. *International Journal of Industrial Ergonomics*, 16, 191–200.

Grönqvist, R., Chang, W.R., Courtney, T.K., Leamon, T.B., Redfern, M.S., and Strandberg, L., 2001c, Measurement of slipperiness: fundamental concepts and definitions. *Ergonomics*, 44, 1102–1117.

Grönqvist, R., Hirvonen, M., and Matz, S., 2001a, Walking safety and contact time related variation in shoe-floor traction. *Proceedings of the International Conference on Computer-Aided Ergonomics and Safety*, July 29–August 1, Maui, Hawaii.

Grönqvist, R., Hirvonen, M., and Rajamäki, E., 2001b, Development of a portable test device for assessing on-site floor slipperiness: an interim report. *Applied Ergonomics*, 32, 163–171.

Grönqvist, R., Hirvonen, M., and Skyttä, E., 1992, Countermeasures against floor slipperiness in the food industry. In S. Kumar (Ed.) *Advances in industrial ergonomics and safety IV* (London: Taylor & Francis), pp. 989–996.

Grönqvist, R., Hirvonen, M., and Tohv, A., 1999, Evaluation of three portable floor friction testers. *International Journal of Industrial Ergonomics*, 25, 85–95.

Grönqvist, R., Hirvonen, M., and Tuusa, A., 1993, Slipperiness of the shoe-floor interface: comparison of objective and subjective assessments. *Applied Ergonomics*, 24, 258–262.

Grönqvist, R., Roine, J., Järvinen, E., and Korhonen, E., 1989, An apparatus and a method for determining the slip resistance of shoes and floors by simulation of human foot motions. *Ergonomics*, 32, 979–995.

Grönqvist, R., Rahikainen, A., Roine, J., and Korhonen, E., 1988b, Alusten kulkuteiden pintamateriaalien liukkaus (Slipperiness of underfoot surfaces in ships). *Työ ja ihminen (People and Work)*, 2, 159–168 (in Finnish with English summary).

Grönqvist, R., Roine, J., and Korhonen, E., 1988a, Suojajalkineiden pitävyys uusina ja käytettyinä (Slip resistance of new and used safety shoes). *Työ ja ihminen (People and Work)*, 2, 149–158 (in Finnish with English summary).

Grönqvist, R., Roine, J., Korhonen, E., and Rahikainen, A., 1990, Slip resistance versus surface roughness of deck and other underfoot surfaces in ships. *Journal of Occupational Accidents*, 13, 291–302.

Harris, G.W. and Shaw, S.R., 1988, Slip resistance of floors: users' opinions, Tortus instrument readings and roughness measurement. *Journal of Occupational Accidents*, 9, 287–298.

Hirvonen, M., Leskinen, T., Grönqvist, R., and Saario, J., 1994, Detection of near accidents by measurement of horizontal acceleration of the trunk. *International Journal of Industrial Ergonomics*, 14, 307–314.

Hirvonen, M., Leskinen, T., Grönqvist, R., Viikari-Juntura, E., and Riihimäki, H., 1996, Occurrence of sudden movements at work. *Safety Science*, 24, 77–82.

Hsiao, E.T. and Robinovitch, S.N., 1998, Common protective movements govern unexpected falls from standing height. *Journal of Biomechanics*, 31, 1–9.

Hsiao, E.T. and Robinovitch, S.N., 1999, Biomechanical influences on balance recovery by stepping. *Journal of Biomechanics*, 32, 1099–1106.

James, D.I., 1980, A broader look at pedestrian friction. *Rubber Chemistry and Technology*, 53, 512–541.

Jung, K., 1991, Einflussfaktoren auf die Rutschemmung. *Die Berufsgenossenschaft*, Heft 2 (Bielefeld: Erich Schmidt Verlag), pp. 1–5 (in German).

Jung, K. and Riediger, G., 1982, Neuere Entwicklungen zur Prüfung der Rutschemmung von Bodenbelägen. *Die Berufsgenossenschaft*, Heft 6 (Bielefeld: Erich Schmidt Verlag), pp. 1–7 (in German).

Jung, K. and Rütten, A., 1992, Entwicklung eines Verfahrens zur Prüfung der Rutschemmung von Bodenbelägen für Arbeitsräume, Arbeitsbereich und Verkehrswege. *Zentralblatt für Arbeitsmedizin, Arbeitsschutz, Prophylaxe und Ergonomie*, 42 (6), 227–235 (in German with English summary).

Jung, K. and Schenk, H., 1989, Objektivierbarkeit und Genauigkeit des Begehungsverfahrens zur Ermittlung der Rutschemmung von Bodenbelägen. *Zentralblatt für Arbeitsmedizin, Arbeitsschutz, Prophylaxe und Ergonomie*, 39 (8), 221–228 (in German with English summary).

Jung, K. and Schenk, H., 1990, Objektivierbarkeit und Genauigkeit des Begehungsverfahrens zur Ermittlung der Rutschemmung von Schuhen. *Zentralblatt für Arbeitsmedizin, Arbeitsschutz, Prophylaxe und Ergonomie*, 40 (3), 70–78 (in German with English summary).

Kim, I.-J. and Smith, R., 2000, Observation of the floor surface topography changes in pedestrian slip resistance measurements. *International Journal of Industrial Ergonomics*, 26, 581–601.

Kummer, H.W., 1966, Unified theory of rubber and tire friction. *Engineering Research Bulletin B-94* (University Park, PA: The Pennsylvania State University).

Lavender, S.A., Sommerich, C.M., Sudhaker, L.R., and Marras, W.S., 1988, Trunk muscle loading in non-sagittally symmetric postures as a result of sudden unexpected loading conditions. *Proceedings of the Human Factors Society 32nd Annual Meeting*, Vol. 1, The Human Factors Society, Santa Monica, California, pp. 665–669.

Leamon, T.B. and Li, K.-W., 1990, Microslip length and the perception of slipping. Presented at the 23rd International Congress on Occupational Health, September 22–28, Montreal.

Leamon, T.B. and Son, D.H., 1989, The natural history of a microslip. In A. Mital (Ed.) *Advances in Industrial Ergonomics and Safety I* (London: Taylor & Francis), pp. 633–638.

Leclercq, S., 1999, The prevention of slipping accidents: a review and discussion of work related to the methodology of measuring slip resistance. *Safety Science,* 31, 95–125.

Leclercq, S., Tisserand, M., and Saulnier, H., 1995a, Tribological concepts involved in slipping accident analysis. *Ergonomics*, 38, 197–208.

Leclercq, S., Tisserand, M., and Saulnier, H., 1995b, Assessment of slipping resistance of footwear and floor surfaces. Influence of manufacture and utilization of the products. *Ergonomics*, 38, 202–219.

Leclercq, S., Tisserand, M., and Saulnier, H., 1997, Analysis of measurements of slip resistance of soiled surfaces on site. *Applied Ergonomics*, 28, 283–294.

Llewellyn, M.G.A. and Nevola, V.R., 1992, Strategies for walking on low-friction surfaces. In W.A. Lotens and G. Havenith (Eds.) *Proceedings of the Fifth International Conference on Environmental Ergonomics*, Maastricht, The Netherlands, pp. 156–157.

Lloyd, D.G. and Stevenson, M.G., 1992, An investigation of floor surface profile characteristics that will reduce the incidence of slips and falls. *Transactions of Mechanical Engineering, The Institution of Engineers, Australia*, ME 17 (2), 99–105.

Lockhart, T.E., 1997, The ability of elderly people to traverse slippery walking surfaces. In *Proceedings of the Human Factors and Ergonomics Society 41st Annual Meeting*, Vol. 1 (Albuquerque: Human Factors and Ergonomics Society), pp. 125–129.

Lockhart, T.E., Smith, J.L., Woldstad, J.C., and Lee, P.S., 2000a, Effects of musculoskeletal and sensory degradation due to aging on the biomechanics of slips and falls. In *Proceedings of the IEA/HFES Congress*, Vol. 5, Industrial Ergonomics (San Diego: International Ergonomics Association), pp. 83–86.

Lockhart, T.E., Woldstad, J.C., Smith, J.L., and Hsiang, S.M., 2000b, Prediction of falls using a robust definition of slip distance and adjusted required coefficient of friction. In *Proceedings of the IEA/HFES Congress*, Vol. 4, Safety and Health, Aging (San Diego: International Ergonomics Association), pp. 506–509.

Lockhart, T.E., Woldstad, J.C., Smith, J.L., and Ramsey, J.D., 2002, Effects of age related sensory degradation on perception of floor slipperiness and associated slip parameters. *Safety Science*, 40, 689–703.

Manning, D.P. and Jones, C., 1993, A step towards safe walking. *Safety Science*, 16, 207–220.

Manning, D.P. and Jones, C., 1994, The superior slip-resistance of footwear soling compound T66/103. *Safety Science*, 18, 45–60.

Manning, D.P. and Jones, C., 2001, The effect of roughness, floor polish, water, oil and ice on underfoot friction: current safety footwear solings are less slip resistant than microcellular polyurethane. *Applied Ergonomics*, 32, 185–196.

Manning D.P., Jones, C., and Bruce, M., 1983, Improved slip-resistance on oil from surface roughness of footwear. *Rubber Chemistry and Technology*, 56, 703–717.

Manning, D.P., Jones, C., and Bruce, M., 1985, Boots for oily surfaces, *Ergonomics*, 28, 1011–1019.

Manning, D.P., Jones, C., and Bruce, M., 1990, Proof of shoe slip-resistance by a walking traction test, *Journal of Occupational Accidents*, 12, 255–270.

Manning, D.P., Jones, C., and Bruce, M., 1991, A method of ranking the grip of industrial footwear on water wet, oily and icy surfaces, *Safety Science*, 14, 1–12.

Manning, D.P., Jones, C., Rowland, F.J., and Roff, M., 1998, The surface roughness of a rubber soling material determines the coefficient of friction on water-lubricated surfaces. *Journal of Safety Research*, 29, 275–283.

Manning, D.P., Mitchell, R.G., and Blanchfield, L.P., 1984, Body movements and events contributing to accidental and nonaccidental back injuries. *Spine*, 9, 734–739.

Marigold, D.S. and Patla, A.E., 2002, Strategies for dynamic stability during locomotion on a slippery surface: effects of prior experience and knowledge. *Journal of Neurophysiology*, 88, 339–353.

Moore, D.F., 1972, The friction and lubrication of elastomers. In G.V. Raynor (Ed.) *International series of monographs on material science and technology*, Vol. 9 (Oxford: Pergamon Press).

Murray, M.P., 1967, Gait as a total pattern of movement. *American Journal of Physical Medicine*, 46, 290–333.

Myung, R., Smith, J.L., and Leamon, T.B., 1992, Slip distance for slip/fall studies. In S. Kumar (Ed.) *Advances in Industrial Ergonomics and Safety IV* (London: Taylor & Francis), pp. 983–987.

Myung, R., Smith, J.L., and Leamon, T.B., 1993, Subjective assessment of floor slipperiness. *International Journal of Industrial Ergonomics*, 11, 313–319.

Nagata, H., 1989, The methodology of insuring the validity of a slip-resistance meter. In *Proceedings of the International Conference on Safety* (Tokyo: Metropolitan Institute of Technology), pp. 33–38.

Nashner, L.M., 1985, Conceptual and biomechanical models of postural control. In Igarashi and Black (Eds.) *Vestibular and Visual Control of Posture and Locomotor Equilibrium* (Karger: Basel), pp. 1–8.

Pai, Y.-C. and Patton, J., 1997, Center of mass velocity-position predictions for balance control. *Journal of Biomechanics*, 30, 347–354.

Perkins, P.J., 1978, Measurement of slip between the shoe and ground during walking. In C. Anderson and J. Senne (Eds.) *Walkway surfaces: measurement of slip resistance*, ASTM STP 649 (Baltimore: American Society for Testing and Materials), pp. 71–87.

Perkins, P.J. and Wilson, P., 1983, Slip resistance testing of shoes — new developments. *Ergonomics*, 26, 73–82.

Proctor, T.D. and Coleman, V., 1988, Slipping, tripping and falling accidents in Great Britain — present and future. *Journal of Occupational Accidents*, 9, 269–285.

Redfern, M.S. and Bloswick, D., 1997, Slips, trips, and falls. In M. Nordin, G. Andersson and M. Pope (Eds.) *Musculoskeletal Disorders in the Workplace*, Chapter 13 (St. Louis: Mosby-Year Book), pp. 152–166.

Redfern, M., Cham, R., Gielo-Perczak, K., Grönqvist, R., Hirvonen, M., Lanshammar, H., Marpet, M., Pai, Y.-C., and Powers, C., 2001, Biomechanics of slips. *Ergonomics*, 44, 1138–1166.

Redfern, M.S. and Schuman, T., 1994, A model of foot placement during gait. *Journal of Biomechanics*, 27, 1339–1346.

Robinovitch, S.N., Hsiao, E., Kearny, M., and Frenk, V., 1996, Analysis of movement strategies during unexpected falls. *Proceedings of the 20th Annual Meeting of the American Society of Biomechanics*, October 17–19, Atlanta, Georgia.

Rowland, F.J., Jones, C., and Manning, D.P., 1996, Surface roughness of footwear soling materials: relevance to slip-resistance. *Journal of Testing and Evaluation*, 24, 368–376.

Scheil, M. and Windhövel, U., 1994, Instationäre Reibzahlmessung mit dem Messverfahren nach Manning. *Zeitschrift für Arbeitswissenschaft*, 20, 177–181 (in German with English summary).

Skiba, R., Bonefeld, X., and Mellwig, D., 1983, Voraussetzung zur Bestimmung der Gleitsicherheit beim menschlichen Gang. *Zeitschrift für Arbeitswissenschaft*, 9, 227–232 (in German with English summary).

Skiba, R., Wieder, R., and Cziuk, N., 1986, Zum Erkinntniswert von Reibzahlmessung durch Begehen einer neigbaren Ebene. *Kautschuk+Gummi Kunststoffe*, 39, 907–911 (in German with English summary).

Stobbe, T.J. and Plummer, R.W., 1988, Sudden-movement/unexpected loading as a factor in back injuries. In F. Aghazadeh (Ed.) *Trends in Ergonomics/Human Factors V* (Elsevier Science Publishers B.V. North-Holland), pp. 713–720.

Strandberg, L., 1983a, On accident analysis and slip-resistance measurement. *Ergonomics*, 26, 11–32.

Strandberg, L., 1983b, Ergonomics applied to slipping accidents. In T.O. Kvålseth (Ed.) *Ergonomics of Workstation Design*, Chapter 14 (London: Butterworths), pp. 201–208.

Strandberg, L., 1985, The effect of conditions underfoot on falling and overexertion accidents. *Ergonomics*, 28, 131–147.

Strandberg, L., Hildeskog, L., and Ottoson, A.-L., 1985, Footwear friction assessed by walking experiments. *VTIrapport 300 A*, Väg-och trafikinstitutet, Linköping.

Strandberg, L. and Lanshammar, H., 1981, The dynamics of slipping accidents. *Journal of Occupational Accidents*, 3, 153–162.

Swensen, E., Purswell, J., Schlegel, R., and Stanevich, R., 1992, Coefficient of friction and subjective assessment of slippery work surfaces, *Human Factors*, 34, 67–77.

Tang, P.-F., Woollacott, M.H., and Chong, R.K.Y., 1998, Control of reactive balance adjustments in perturbed human walking: roles of proximal and distal postural muscle activity. *Experimental Brain Research*, 119, 141–152.

Tisserand, M., 1985, Progress in the prevention of falls caused by slipping. *Ergonomics*, 28, 1027–1042.

Waller, J.A., 1978, Falls among the elderly — human and environmental factors. *Accident Analysis and Prevention*, 10, 21–33.

Wieder, R., 1988, Experimentelle Untersuchungen über den Einfluss der Oberflächenrauheit auf die Gleitsicherheit beim menschlichen Gang. Fachbereich Sicherheitstechnik der Bergischen Universität — Gesamthochschule Wuppertal, doctoral dissertation (in German).

Winter, D.A., 1991, *The biomechanics and motor control of human gait: normal, elderly, and pathological*, 2nd ed. (Waterloo, Ontario: University of Waterloo).

Winter, D.A., 1995, *A.B.C.: anatomy, biomechanics and control of balance during standing and walking* (Waterloo, Ontario: University of Waterloo).

Yoshioka, M., Ono, H., Kawamura, S., and Miyaki, M., 1978, On slipperiness of building floors — fundamental investigation for scaling of slipperiness. In *Report of the Research Laboratory of Engineering Materials, No. 3* (Tokyo: Tokyo Institute of Technology), pp. 129–134.

Yoshioka, M., Ono, H., Shinohara, M., Kawamura, S., Miyaki, M., and Kawata, A., 1979, Slipperiness of building floors. In *Report of the Research Laboratory of Engineering Materials, No. 4* (Tokyo: Tokyo Institute of Technology), pp. 140–157.

Chapter 3

Steps and Stairs

Mike Roys

CONTENTS

3.1 Introduction

As an architectural focal point, stairs are often an elegant feature within buildings. At the same time, steps and stairs are highly functional, providing quick, efficient, and permanent means of access between levels. Stairs vary in size in both their overall scale and in dimensions such as the rise and going (i.e., tread depth) of individual steps (Figure 3.1). Stairs vary in shape, with straight, spiral, doglegged, alternating, and geometric designs among those found. Stairs vary in the materials from which they are constructed, concrete, wood, metals, stone, and even glass, for example. They vary in location, and thus function, such as those between storeys in a home, stairs in an office, stairs in a sports stadium or other public building, steps down into a swimming pool, or the assortment of steps and stairs found in an outdoor urban landscape. Entire books (e.g., Templer 1992) have been written to try to describe this variety and the ways in which people use the range of stairs and steps encountered in everyday life. This chapter discusses the design of stairs, focussing on those aspects of most importance for falls and their prevention.

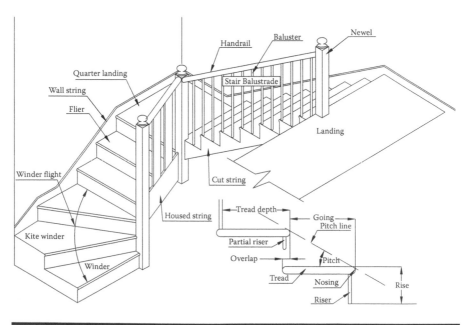

Figure 3.1   Definitions for rise, going, and pitch.

Table 3.1 The Incidence of Stair Injuries by Location in the United Kingdom

Age Group	Fatal[a]	Home Injuries		Leisure Injuries[b]	Workplace Injuries[c]	Total
		Serious Injuries[b]				
0–4	5	45,706		4,097	—	103,579
5–14		31,074		22,697	—	—
15–64	150	169,163		57,850	1176	228,294
65–74	105	18,436		8,175	—	26,716
75+	240	25,697		13,223	—	39,160
Total	500	290,076		106,042	1176	397,749

[a] Estimates based on data for 1995 from the Office for National Statistics (ONS).

[b] 1999 Home and Leisure Accident Surveillance System data. (From DTI, 2001. *23rd annual report of the home and leisure accident surveillance system — 1999 data.* (London: Department of Trade and Industry), DTI URN 01/32.)

[c] Average over 5 years 1994/1995 to 1998/1999. (From HELA 2000 reports.)

3.2 The Extent of the Problem

Every year in the United Kingdom, 290,000 individuals have a serious injury in their home involving stairs, and at least 500–600 people die because of a fall in this location. In addition, approximately 100,000 serious injuries happen on stairs during leisure activities outside the home, with a further 1000 falls occurring on stairs in the workplace (Table 3.1). This equates to an incidence rate of 1 in 150 individuals injured each year through falling on stairs, or an injury on stairs happening every 80 seconds in the United Kingdom. Similar incidence rates are found in other countries. Interpretation of these statistics needs to bear in mind the small proportion of our time spent on stairs compared with other locations.

3.3 Gait on Stairs

Walking on stairs is a process learned at a very early age, with most children able to ascend stairs in an upright posture by the age of two, or earlier if assisted by a parent, and able to descend stairs without sliding or shuffling soon after. Walking gait on stairs is, however, very different from walking on the level (see Chapter 2). Typical gait patterns for ascent and descent are described in the following sections.

3.3.1 Ascent

In ascent, the stair user first assesses the flight visually on approach. The leading foot is then raised in front of the user, beyond the height of the first step, allowing the foot to be placed on the tread, toes first.

The user's weight is transferred to this forward foot, while simultaneously pushing off the floor with the rear foot and lowering the heel of the front foot, until it makes contact with the tread. Where the step has a going that is smaller than the user's foot, it may be impossible to place the heel on the tread without turning the foot away from the direction of travel. Most able-bodied users are capable of sustaining their weight on the ball of the foot, however, and can manage to ascend the stair without the heel making contact with the step.

The rear foot then swings past the leading foot, travelling over the nosings of two consecutive steps, to rest on the next step up the flight in a similar manner to the initial step.

This process is repeated throughout the flight until the user alights on the landing at the top. By the time the third step has been taken, the user has determined the expected rise of the steps through proprioceptive feedback and can ascend the flight without looking at the steps. During this stage, the toes of the rear foot swing past the nosings with only a few millimetres to spare, and the foot is placed on the next step almost horizontally.

3.3.2 Descent

In descent, the stair user again first assesses the flight visually for hazards or any unexpected problems, while placing the leading foot near to the edge of the landing. The user then proceeds to initiate the descent, allowing the rear foot to swing past the landing nosing into the air above the next step.

At the same time, the user bends the knee of the leg supporting his or her weight, with the heel of the foot rising off the landing surface. At this point, the user is committed to taking the first step down the flight.

The toes of the other foot, suspended over the first step, are pointed down toward the step tread. The foot is pulled slightly back toward the body and, on contact with the step, the weight is transferred on to this new leading foot. The descent is controlled as the heel of the leading foot comes to rest on the tread.

The rear foot is then lifted off the step above, swinging within millimetres of the next two nosings and the pattern repeated. The user may take another couple of glances at the steps on descent, but will normally, again through proprioceptive feedback, descend the rest of the flight without looking or thinking about what they are doing.

3.3.3 Gait Disruption

The gait patterns described in Section 3.3.1 and Section 3.3.2 are those found on straight stairs and occur throughout the flight, providing the steps are consistent and adequate. Differences in dimensions of steps within stairs may interfere with a user's gait, sometimes resulting in a stumble or fall. Other factors that can affect the user and his or her gait include objects being left on the steps, variation in lighting and shadows, distractions in the visual field around the stairs, or the effect of other people using the stairs at the same time.

Climbing and descending stairs requires physical fitness and coordination. A decline in these, as may occur with aging or illness (see Chapter 5), increases the vulnerability of users to falling. Not only is a user's ability affected by his or her age, weight, height, strength, and stamina, but also by what else he or she might be doing at the time. For example, individuals sometimes climb or descend stairs carrying children or other objects; they may do so reading, eating, or perhaps nowadays using a mobile phone; or they may be distracted, running, or turning round on the stairs to travel in the opposite direction.

Although individual and behavioural influences, such as those described, are contributors to fall events on stairs, poor design and construction remain important primary factors. The four main problems that are likely to be encountered are: inconsistent step dimensions, inappropriate step sizes, inadequate handrails, and poor step visibility. Many stair falls can be related to one or more of these four inadequacies and the remaining sections of this chapter consider each of these in detail.

3.4 Inconsistent Step Dimensions

3.4.1 The Problem

It is known from observation and video analysis (Cohen 2000) that during normal gait on stairs, the foot passes over and very close to two nosings with every step the user makes. Proprioceptive feedback from the lower limbs enables the step size within a flight to be judged quickly, allowing the user to make a series of identical steps in rapid succession. This process takes just fractions of a second, and the measure of the whole flight may be determined in the first step, particularly if the flight is familiar. Even where the steps are unfamiliar, the user is likely to have established the size of the steps within the first three steps. Occasionally, however, the steps that comprise a flight are not as consistent as the initial proprioceptive feedback to the user might suggest. Inconsistencies due to poor construction, wear on the stair surface, or damage, may lead to differences of perhaps 5–25 mm in rise and going of steps within a stair. Even small variations may affect the user.

3.4.2 Differences in Rise

When designing stairs for a building, the first process is usually to calculate the floor-to-floor height from the building plans or from direct measurement on site. A value for the rise is then calculated by dividing this dimension into equal proportions so that the final rise, going, and pitch will be appropriate for the type of use, as determined by standards and building codes. For example, in England, maximum values for the rise

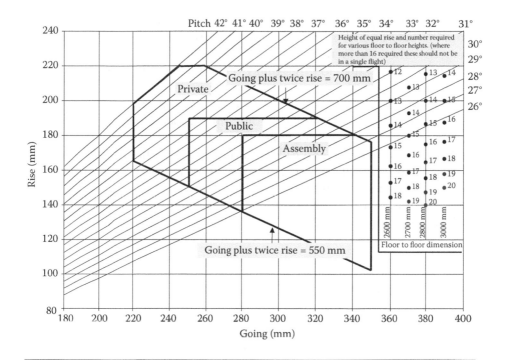

Figure 3.2   Acceptable dimensions for stairs in England and Wales.

(220 mm), pitch (42°) and a minimum value for going (220 mm) are recommended for domestic stairs (DETR 1997). The British Standard for stairs BS5395-1 (BSI 2000) also suggests a maximum going (350 mm) and a minimum rise (100 mm). Comfort is allowed for by recommending suitable values for a combination of rise and going (550 mm < going + [2 × rise] < 700 mm). Thus, if the floor to floor height is expected to be 2.6 m, a stair with 13 steps can be created, with a rise of 200 mm, a going of 223 mm and a pitch just below 42° (see Figure 3.2). In the United States, a movement is currently under way to adopt a single standard for stairs, irrespective of location, with a maximum rise of 7 inches (180 mm) and a minimum going of 11 inches (280 mm). This has become known as the 7-11 rule.

When prefabricated stairs are delivered to site, it is sometimes found that the actual floor-to-floor dimension is slightly different from that calculated based on the building plans. In such situations, the discrepancy is accommodated by allowing the bottom or, occasionally, the top rise to be slightly different from the rest of the flight. When steps are cast *in situ*, often from concrete, or when poor detailing is made to prefabricated strings, the step rise can vary throughout the flight. Guidelines for step-to-step consistency are given in British Standard BS5606 (BSI 1990), which suggests that a typical cast *in situ* concrete flight should have a maximum variation between step rises of no more than 4–6 mm. If this variation is exceeded, the risk increases that the user will trip on the nosing of the next tread, due to the close proximity of the foot travel to the step surface, as already described.

In ascent, this may cause the user to fall forward onto the steps above, sometimes resulting in injuries to the arms, upper torso, and potentially the head. The length of such a fall is usually limited to the user's height, and thus these injuries tend to be minor, although influenced by the type of finish and material used to cover the steps. In descent, the user's foot may make contact with the next tread earlier or later than expected. Normally this will have little effect, but on occasion may cause the user to lose his or her balance, possibly falling down the remaining steps in the flight.

3.4.3 Differences in Going

Differences in going between steps can occur in a similar way to differences in rise. In addition, it is common in some countries to find the nosing missing on the top step of a flight. This occurs where prefabricated steps are built so that the top fabricated step in the flight is designed to attach to the floor structure one rise down from the landing (Figure 3.3). It becomes necessary, therefore, to attach a separate nosing to the landing to form a consistent going throughout the flight. It is common, however, for the installation of this separate nosing to be overlooked.

Where the going varies between two steps, so that a following step encountered in descent is smaller than the first, the user may overstep the tread, leading to a fall. In such circumstances, the overstepped foot usually slides over the edge of the nosing causing the user to fall backward

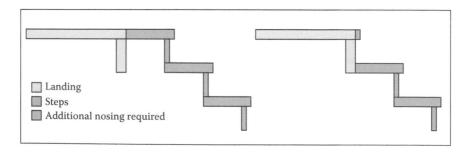

Landing
Steps
Additional nosing required

Figure 3.3   Alternative methods of attaching a flight of steps to an existing landing.

and injure his or her legs, back, and possibly the head. This can be particularly serious for older users who may suffer upper leg or hip fractures (e.g., broken neck of femur-type injuries). Occasionally, the user may get the heel of his or her shoe caught on the nosing or may stumble forward down the flight, resulting in injuries that are more serious to the head, arms, and torso.

In ascent, a smaller going leading to a larger going might form a tripping hazard, in a similar way to an inconsistent rise, leading the user to fall forward on the stair. In addition, in ascent, a larger going leading to a smaller going may cause the user to under step, resulting in the foot sliding off the nosing, again causing the user to stumble forward. The potential exists for injuries to the arms, torso, and possibly the head in both these situations.

It may be observed from these sections that consistent step geometry is essential for safety on stairs.

3.5 Inappropriate Step Sizes

3.5.1 The Problem

Some stairs are easier to use than others, some are more comfortable, and some are safer. Suggestions have been made throughout the ages as to the best design for stairs. In the first century A.D., a fixed ratio of 3 to 4 (rise to going) was recommended (Templer 1992). By 1570, limits for the rise between 114–173 mm and for the going between 305–457 mm were defined. It is interesting to note that these recommendations remain appropriate today and would lead to stairs suitable for all ambulant users. By 1686, stairs that were more generous were being recommended, with a maximum rise of 127 mm and minimum going no less than 381 mm.

Around this time, attempts were also made to relate the combination of suitable rise and going dimensions to the length of a user's stride. One such attempt by François Blondel is still used in some contemporary building standards throughout Europe (CEN 2002). Everybody has a slightly different stride length on stairs, however, and, because of this, the step geometry that will suit one person will be less satisfactory for another. It is, therefore, impossible to design stairs that are optimum for all users, making compromise necessary. Thus, most codes and standards suggest minimum or maximum values for stair dimensions. In England, a maximum value is set for rise and a minimum value for going, with different values for these according to the nature of the building in which the stairway is to be installed.

3.5.2 Minimum Going

In both ascent and decent it is usually the forefoot (phalangeal/metatarsal region) that makes initial contact with each step. In descent, this can be a problem if the going is small. Ideally, the going should be large enough to allow the whole foot to fit on the step (Roys 2001). For an average male, this would mean the going would have to be at least 300 mm, including a 30-mm correction for shoes. In practice, however, stairs in England are built close to the minimum allowed dimensions. This means that in domestic buildings, most stairs are built with goings close to 220 mm, with 95% of all domestic stairs having a going less than 240 mm. For assembly buildings, where a large number of people gather, many of whom may be unfamiliar with the environment, the minimum going is 280 mm. For all other buildings, the minimum going is 250 mm. If the going is smaller than the user's foot, which is true for nearly all domestic stairs in England, Europe, Japan, and probably elsewhere, the risk increases for overstepping during descent. This is perhaps the most common type of fall initiating event on stairs in the home.

To accommodate small steps, individuals often turn their feet outward away from the direction of travel. Recent research (Roys 2002a) suggests that as the going is decreased below 250 mm, the angle of the foot increases from an average of 15° to 30° or more. The angle continues to increase with decreasing going, so that the foot hangs over the tread by a consistent amount, approximately 80 mm on average; however, some individuals continue to keep their feet parallel to the direction of travel no matter how small the going. For these persons, as much as 50% of the foot might overhang the nosing on stairs with a going at the lower end of the range (Figure 3.4).

If 30% or more of the foot extends beyond the tread during descent, it is likely that the downward slope of the foot on contact with the step will continue after contact, and the leading foot may slide over the nosing. To correct the ensuing imbalance, either the rear leg will be flexed, or the rear foot will swing forward to try to bring the centre of mass of the user back within normal walking limits. If successful, the user may prevent a serious fall, finding himself or herself sitting on the steps, bruising little more than his or her upper legs and ego. If unsuccessful, the user may suffer leg injuries, or may even fall forward, with potential for injury to their upper body, arms, and head.

In practice, nearly 50% of users have 30% or more of their foot beyond the edge of the nosing when the going is 250 mm. With a going of 300 mm, only 10% of users have such an overhang. On these grounds, it is suggested that a going of 250 mm for main domestic stairs should be an absolute minimum. In the United States, the 7-11 recommendation

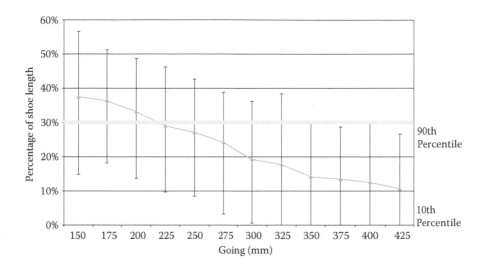

Figure 3.4   Percentage overhang of users' feet by going size.

applies to domestic stairs, that is to say a minimum going of 11 inches (280 mm). Until goings of at least this size are the norm, overstepping on domestic stairs will remain a common cause of falls. Recent research suggests that goings as large as 450 mm may be appropriate, when combined with a suitable rise (Roys 2002a).

3.5.3 Maximum Rise

Achieving an appropriate sized rise is perhaps less of a problem than providing a suitable sized going. It appears that most people are comfortable with any rise between 150–200 mm. If the rise is smaller than 100 mm, the steps may become a trip hazard in ascent, and the user may feel that excessive effort is required to achieve vertical progress. Research by Lehmann and Englemann (1933), cited by Templer (1992), which studied energy use on stairs, concluded that smaller rises should be combined with smaller goings so that the difference between them is always 120 mm. Templer notes, however, that the speed of travel was fixed for each rise and going combination, which is not the way users will normally use stairs. When allowing users to vary their travelling speed, Templer and colleagues were able to demonstrate that higher rises with smaller goings cause the users to expend no more energy than stairs with smaller rises and larger goings.

When descending stairs, the size of the rise determines both the angle (extension) with which the foot approaches the step (the larger the rise the greater the angle) and the extent of the user's command over the descent. If the user is frail, elderly, or a young child, then a stair with a smaller rise between 100–180 mm, may make the whole flight easier to manage. When the rise exceeds 200 mm, the risk increases for these users because the descent between steps may be too great for them to control. Some older users prefer a larger rise, however, if it means fewer steps are needed to reach the same destination, because the risk of a fall is at its greatest each time a step is taken. For younger people and athletic types, taking one step at a time is just too slow, and thus such individuals will often move up or down stairs at a rate of two or even three steps at a time, depending on the rise.

For each step in ascent, the user has to lift their whole body weight up the height of one rise for every step. As the rise increases, certainly beyond 200 mm, this becomes more difficult for some users. For such people, handrails are important to help them make the climb successfully.

3.6 Inadequate Handrails

3.6.1 The Problem

Handrails, though not always used, have three basic functions. First, they act as a support to the user. Second, they provide guidance to the user, helping them determine the direction and pitch of the stair, where the flight starts, and where it ends. The third purpose of handrails is in aiding correction of a loss of balance when this occurs, an important function in terms of preventing falls. Handrails are used differently during ascent and descent. In ascent, the user may grasp the handrail to assist lifting their body weight up each step. This becomes more necessary as the user becomes elderly, frail, or when the rise is increased beyond 200 mm. In descent, the handrail is often only lightly touched or the hand is allowed to slide down the full length of the handrail as the user descends.

When a loss of balance occurs, Maki et al. (1998) found that in descent, users are able to catch hold of a handrail and generate a significant stabilising force within just one second of balance loss onset; however, the handrail does have to be positioned appropriately so that the hand can reach it quickly. In addition, the handrail needs to be adequately sized so that the hand can be clasped around the whole rail in a power grip, providing the handhold needed to stop the fall.

Because a fall can commence on any step within a flight of stairs, it is important that the handrail be within reach throughout the whole of the flight, regardless of the number of steps. Recent research (Roys 2002b) has demonstrated that the risk of having a serious injury on stairs with no handrails (i.e., a flight between two walls) is twice that of a flight with two handrails. Incidentally, if the handrail and protective barrier (either guarding or a wall) is removed from one side of the flight, the risk is doubled again, even if a handrail is on the other side.

3.6.2 Handrail Height

Considerable attention has been devoted to determining the appropriate height of handrails (measured as the vertical distance between the top of the handrail and the pitch line — the pitch line is the nominal line connecting the nosings of a flight of steps; see Figure 3.1). The most important criterion for the height of a handrail is that it should be possible to grasp it quickly in the event of a fall. Maki et al. (1984) found a handrail height between 900–1000 mm (36–40 inches) to be the most effective for counteracting the forces present in descent falls, and a recommended handrail height of 900 mm is now found in many national building standards.

It should be noted that a handrail, to be effective in a fall situation, must be rigid and secure. Most handrails are made from solid wood or metal, sometimes with a plastic coating, and thus this is not usually a problem. Handrails constructed from rope or other flexible materials are less useful for preventing falls.

3.6.3 Handrail Shape

The shape of a handrail affects the force that can be exerted by the user's hand, with Maki (1985) again influential in researching this area. The two most common forms of handrail found on domestic stairs in England are the pig's ear and traditional profiles (Figure 3.5). Unfortunately, both of these are poor at affording users a power grip.

Figure 3.5   Profiles of pig's ear and traditional handrails.

Reinforcing their use is that both the pig's ear and traditional handrails are considered aesthetically pleasing, whereas alternatives are often thought to be too "institutional" for domestic use. Circular handrails with a diameter between 32–50 mm offer the best combination of size and shape for enabling a strong grasp in normal use. Oval handrails with comparable circumferences (100–150 mm) offer similar advantages. It is pleasing to note that major retailers in the United Kingdom are now offering circular handrails as an alternative to the pig's ear and traditional styles.

The frictional properties of the handrail material are also important for the purposes of fall prevention. If the handrail is too rough, then the user is less likely to want to use it in normal circumstances and, in the event of a fall, the handrail may cause abrasions or lacerations to the hand. If the handrail is too smooth, then the user may not be able to prevent a fall because his or her hand may simply slip along the rail.

3.7 Poor Step Visibility

3.7.1 The Problem

In the description of gait on stairs at the beginning of this chapter (Section 3.3), it was described how users precede their ascent or descent with a visual assessment on the approach to the stairs. This usually happens automatically, without the user needing to think about it. As with crossing a road, it is important when entering a hazardous environment to determine the potential risks, assess the likelihood of harm and determine how best to proceed. Stairs are no different. The visual check allows step spacing to be assessed and potential obstacles to be identified. If something is unusual about the stairs (e.g., if the covering is loose, perhaps some of the steps missing, or a heap of bed sheets have been left on the stairs), the user may decide not to attempt to negotiate the flight, not at least without moving the sheets in the latter instance. Often though, the user will proceed, exercising a greater degree of care than might usually be the case. Good visibility, allowing hazards to be detected, cannot make dangerous stairs safe, but may allow them to be negotiated with caution.

3.7.2 Adequate Lighting

For these reasons, provision of adequate lighting on stairs is important. For indoor stairs, artificial lighting is essential, with two-way lighting switches provided at the top and bottom of a flight. Where multiple flights exist in a stair, lighting switches may be needed at all landings leading to further rooms. In the United Kingdom, the CIBSE lighting codes

(CIBSE 1994), and now the British Standard for stairs (BSI 2000), recommend a minimum lighting level on stairs of 100 lux. This may be higher than is normally achieved on domestic stairs but is a good target to aim for.

A behavioural aspect of the use of lighting is that as users become familiar with particular stairs, as found in their own home for example, they may on occasion proceed with little or no illumination (Hill et al. 2000). This serves to highlight how people can underestimate the risks associated with stairs, and how sometimes they are used without the care and attention they deserve.

Natural light is often also considered desirable for stairs. This perhaps derives from architectural custom and practice, which views stairs as a feature or focal point in a building and, as such, should be lit in a natural and pleasing way; however, care is necessary with the provision of natural lighting because this may lead to heavy shadows in some areas. Natural light can also lead to problems with glare if the window is placed in such a way that bright and direct sunlight shines into the eyes of the user. Under such circumstances, an individual might not be able to see any hazards on the stairs, or may wrongly estimate the position of the next step. Windows provided to light stairs naturally should, therefore, be in positions that are perpendicular to the direction of travel. With regard to other aspects of the use and maintenance of windows, it is important they are located in a manner that allows curtains to be drawn and cleaning to take place safely.

3.7.3 Nosing Contrast

One final measure that should be considered for stairs is contrasting nosings. When a stair is well lit, it is usually quite easy to see the edge of each tread. When the lighting is poor or when the stair covering has a detailed pattern, however, the position of the tread edges may become camouflaged and difficult to discern (see Chapter 4). The same may be true for some types of stair material, particularly untreated glass, where no part of the tread may be clearly visible.

For those with poor eyesight, even in good lighting conditions, the user may have difficulty determining the edge of the treads no matter what material is used. High contrast between the nosing and the rest of the tread is therefore desirable (Figure 3.6). This can be achieved by installing a contrasting strip, extending onto the tread by around 50 mm. On nondomestic stairs, this is often brought about by fitting proprietary nosings, with an inset material in a colour that contrasts well with the carpet or other floor finish. Such nosings may, however, lead to other hazards if the going is small or if the nosing is not fastened properly. An alternative solution can be to mark the edge of treads with a painted

Figure 3.6   High contrast between nosing and step edge.

stripe. A painted stripe can even be applied to domestic carpeted stairs, although some householders may regard this as unsightly.

3.8 Conclusions

Safe stairs have steps of a consistent and appropriate size, with a variation of no more than 5 mm between consecutive rise or going dimensions, rise values preferably between 150–200 mm, and with as large a going as possible. Goings as great as 450 mm are acceptable to most people, when combined with an appropriate-sized rise. In addition, continuous handrails should be provided on at least one, but preferably both, sides of stairs throughout the whole flight, even if the flight is a single step. Handrails should be located 900 mm above the nosing line of the flight, and be of a size and shape so that they may be grasped with a power grip should the user fall. Finally, adequate lighting and nosing contrast should be provided to ensure that, even if the other requirements are not met, the user still has a chance of making it up or down the flight without falling. In the time it has taken you to read this chapter, at least 10 people will have had a serious fall on their stairs at home, just in the United Kingdom. Good stair design has an important role to play in preventing these incidents.

References

British Standards Institute (BSI), 1990. *BS 5606: guide to accuracy in buildings* (London: British Standards Institute).

British Standards Institute (BSI), 2000. *BS 5395-1:stairs, ladders and walkways — Part 1: code of practice for the design, construction and maintenance of straight stairs and winders* (London: British Standards Institute).

Chartered Institution of Building Services Engineers (CIBSE), 1994. *Code for interior lighting* (London: CIBSE).

Cohen H. H., 2000. A field study of stair descent. *Ergonomics in Design*, 8 (2), 11–15.

Comité Européen de Normalisation (CEN), 2002. EN ISO 14122-3: *safety of machinery — permanent means of access to machinery — Part 3: stairs, stepladders and guard-rails* (London: British Standards Institute).

Department of the Environment, Transport and the Regions (DETR), 1997. *Building regulations for England and Wales. Approved Document K: protection from falling, collision and impact*, 1998 Edition (London: HMSO).

Department of Trade and Industry (DTI), 2001. *23rd annual report of the home and leisure accident surveillance system — 1999 data.* (London: Department of Trade and Industry), DTI URN 01/32.

Health and Safety Executive/Local Authority Enforcement Liason Committee (HELA), 2000a. *Key fact sheet on injuries within the retail distribution industry reported to local authorities 1994/1995 to 1998/1999* (London: Health and Safety Commission).

HELA, 2000b. *Key fact sheet on injuries within the hotel and catering industry reported to local authorities 1994/1995 to 1998/1999* (London: Health and Safety Commission).

HELA, 2000c. *Key fact sheet on injuries within the consumer/leisure service industry reported to local authorities 1994/1995 to 1998/1999* (London: Health and Safety Commission).

HELA, 2000d. *Key fact sheet on injuries within residential care homes reported to local authorities 1994/1995 to 1998/1999* (London: Health and Safety Commission).

HELA, 2000e. *Key fact sheet on injuries within the wholesale distribution industry reported to local authorities 1994/1995 to 1998/1999* (London: Health and Safety Commission).

HELA, 2000f. *Key fact sheet on injuries within the office-based industries reported to local authorities 1994/1995 to 1998/1999* (London: Health and Safety Commission).

Hill L. D., Haslam R. A., Howarth P. A., Brooke-Wavell K., and Sloane J. E., 2000. *Safety of older people on stairs: behavioural factors* (London: Department of Trade and Industry), DTI ref 00/788.

Maki B. E., 1985. *Influence of handrail shape, size and surface texture on the ability of young and elderly users to generate stabilizing forces and moments* (Ottawa: National Research Council of Canada).

Maki B. E., Bartlett S. A., and Fernie G. R., 1984. Influences of handrail height and stairway slope on the ability of young and elderly users to generate stabilising forces and moments. *Human Factors*, 26, 355–359.

Maki B. E, Perry S. D., and McIlroy W. E., 1998. Efficacy of handrails in preventing stairway falls: a new experimental approach. *Safety Science*, 28, 189–206.

Office for National Statistics (ONS), 1997. Twentieth century mortality — 95 years of mortality data in England and Wales by age, sex, year and underlying cause.

Roys M. S., 2001. Serious stair injuries can be prevented by improved stair design. *Applied Ergonomics*, 32, 135–139.

Roys M. S., 2002a. The effect of changing going on stair safety. In: *Proceedings 6th World Conference on Injury Prevention and Control, Montreal, 12–15 May 2002*, pp. 604–605.

Roys M. S., 2002b. The risk associated with various stair parameters. In: *Proceedings 6th World Conference on Injury Prevention and Control, Montreal, 12–15 May 2002*, pp. 602–603.

Templer J., 1992. *The staircase: studies of hazards, falls and safer design* (Massachusetts: MIT Press).

Chapter 4

Role of Vision in Falls

Peter Howarth

CONTENTS

4.1 Introduction

The extent to which most of us depend on vision for preventing falls can be experienced by closing our eyes, walking around, and seeing how long it is before we fall over something. The detection of objects, obstacles and changes in the walking surface is not, however, the only role that is played by vision when navigating our environment. As well as telling us what is in the world around us, vision, in conjunction with our other senses, also informs us of the whereabouts of our body in space.

Our sense of balance arises mainly from information provided by the somatosensory (i.e., monitoring of muscle activity) and the vestibular (i.e., providing information about movement) systems. In addition, we use other information, such as noise, to produce a consistent internal model of the three-dimensional world and our position within it. Vision makes an important contribution to this internal model. Although the vestibular and somatosensory systems are generally considered crucial for balance (e.g., if we close our eyes, we can still stand upright), the visual system augments these. The closing of the eyes has been reported to increase postural instability (Brooke-Wavell et al. 2002), and although most people can stand on one leg without difficulty, if we close our eyes while doing so it is not long before we overbalance.

Although studies of visual risk factors for falls (e.g., Lord and Dayhew 2001) have generally concentrated on specific attributes, such as visual acuity, additional issues need to be considered. These range from the involvement of sight in locomotion, to how the wearing of spectacles affects risk of falling. This chapter gives broad consideration to aspects of vision that are relevant to falls, including a discussion of lighting and camouflage, as well as the more fundamental topics of vision and depth perception.

4.2 Vision and Conspicuity

The role of vision in preventing slips and trips during locomotion is twofold. First, it exposes the presence of objects, allowing them to be evaded. Second, it reveals changes in the environment, such as a slope or a step, which necessitate a change in locomotive action.

The detection of an object depends upon a characteristic referred to as its "conspicuity." This can be thought of as the totality of the attributes that make an object stand out from its background. We can identify two different aspects to conspicuity. The first is the comparison between the physical characteristics of the object and those of its background; the second is whether the object would be noticed psychologically, and this is likely to be related to its importance.

Figure 4.1 **The ginger cat is difficult to see against a ginger carpet. (From Brace, C.L., Haslam, R.A., Brooke-Wavell, K., and Howarth, P.A. (2003)** *The contribution of behaviour to falls among older people in and around the home.* **Loughborough University for Department of Trade and Industry, Leicestershire (http://www.lboro.ac.uk/departments/hu/groups/hseu/publications).**

Much conspicuity research has been undertaken in the field of transport; following an accident, many vehicle drivers justify their actions with the phrase "I looked but I just didn't see him." Motorcyclists are particularly at risk from this type of problem, and by studying why these accidents happen we can extrapolate from the road situation to the home environment. An analysis of the visual stimulus, the motorcycle, will generally indicate it was supra-threshold in terms of its detectability (i.e., its physical characteristics such as its size), but it still was not seen. An alternative way of thinking about this is that it did not reach consciousness because it was "overlooked." One theory of why this happens is that a driver at a junction who takes a glance down the road may be looking for other cars, which (subconsciously) pose a threat. A motorcycle, coming head on, has a small profile and so is not (subconsciously) much of a danger. Thus, it fails to attract the driver's attention.

In determining what makes something stand out from its background, in the physical sense, we can gain insights by seeing how to make an object disappear into the background (i.e., camouflage). Figure 4.1 depicts a ginger cat sitting on a ginger carpet (a real situation encountered during our research into risk factors for falling among older people [Brace et al. 2003]). In this instance, the outline of the cat is not particularly clear. This is because the colours of the object and the surround are similar, and thus the outline of the cat is not distinct. The importance of the breaking up of outlines can be seen in the camouflaging of military vehicles; the vehicle colours may not match the background precisely, but the "blotchy" pattern still make them difficult to pick out.

Adopting the opposite strategy will enhance the conspicuity of an object, and the difference in conspicuity of two cats — one with an outline that is much more distinct than that of the other, is apparent in Figure 4.2.

Figure 4.2   Emma, the black cat, is more conspicuous than Portia, the tortoiseshell cat, in these surroundings.

We should not forget here that conspicuity is not a property of the object, but instead is a property of the object in a particular context. In Figure 4.1 and Figure 4.2, it is not only the colour and pattern of the cat markings that contribute to the low conspicuity, but also the pattern on the carpet. The object outline is broken up by the carpet pattern — in places the similar colour of the carpet and the cat cause one to blend in with the other. In the United Kingdom, a police motorcycle with orange and white stripes is conspicuous because this pattern of the colours does not occur frequently in an urban environment, and thus they "stand out."

These are examples of objects in particular environments. If conspicuity is different in different environments then it follows that manipulation of the environment can increase conspicuity, and this issue is discussed further in Section 4.5 of this chapter.

4.3 Visual Ability

Visual performance changes with age for two main reasons other than disease. First, the ability to focus on near objects declines. Functionally, when people are no longer able to focus on a near object, such as a book or a newspaper, they purchase "reading spectacles" to improve their near vision. In general this will happen when people are in their forties, and this functional change is termed "presbyopia." Second, with age, the lens yellows and becomes "cloudy," and this is termed a "cataract." The consequence of a cataract is increased intraocular scatter, which manifests itself as a reduction in the apparent contrast of objects. This cannot be corrected optically and, when visual performance has declined to an unacceptable level, the person will have the cataract removed and may subsequently need to wear spectacles for distance as well as for near vision.

4.3.1 Working Distances

4.3.1.1 Distance Vision

The role of distance vision in fall accidents is likely to be minor. In this context, people can cope with degraded distance vision (e.g., that experienced by a myope [a short-sighted person] not wearing his or her spectacles) with little problem. This person can still see objects at different distances, albeit they appear blurred. For a person with normal vision this is like watching a film that is not correctly focussed. Although balance may be affected slightly, it is unlikely to be compromised seriously.

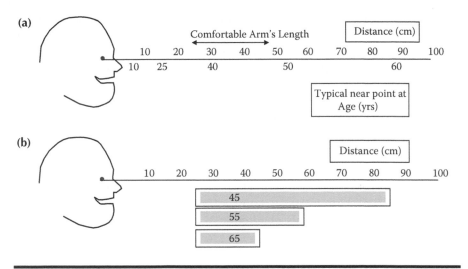

Figure 4.3 **(a) The change in near point with age when no near-vision spectacles are worn. (b) The range of clear vision with age when appropriate near-vision spectacles are worn.**

4.3.1.2 Intermediate Vision

As presbyopia increases (with age), a person's "near point" recedes, as depicted in Figure 4.3a. In addition, the closest they can view something for an extended amount of time without experiencing adverse symptoms also recedes. If a person needs to view an object a little further away than arm's length — which is the case for certain occupations, and may apply to VDU viewing — then they may have a correction for this "intermediate" distance. Such a correction would usually nowadays be in the form of multifocal lenses, but in the past, "trifocals" would have been worn to provide adequate correction for distance, intermediate, and near vision.

4.3.1.3 Near Vision

For most people, because of presbyopia, it is inevitable that near spectacles will become necessary later in life to allow them to continue to see detail, such as print, at close distances. Presbyopia can be explained as follows. A young person can increase the power of his or her eyes naturally through a process called *accommodation*. This increase in power is needed to allow close objects to be seen clearly, in the same way that a camera needs to be focussed differently for distant and near objects. As the eyes age, however, they slowly lose their ability to accommodate.

Near-vision spectacles then perform the function of focussing the light coming from a near object. The older one becomes, the higher the power needed to compensate for the increasing loss of accommodation. Unfortunately, the improvement in near vision comes at a price. The negative consequence of the wearing of reading spectacles is that only a limited range of clear vision is possible while wearing them (see Figure 4.3b), and distant objects are no longer seen clearly.

4.3.2 Spectacles

Single vision lenses have the same optical power over the whole of their surface, as illustrated in Figure 4.4, and are generally used for either distance or near vision. Like most lenses, these will provide both focussing and either magnification (positive lens) or minification (negative lens), the amount of which depends upon the lens power*. Distance spectacles are used to overcome a refractive error (a mismatch between the length of the eye and its optical power), which is essentially independent of age for adults. The accommodative ability of the young means that, even if they need a distance correction, they will not need separate near (reading) spectacles. The older person, however, will need these aids and, as the decrease in accommodative ability is age-related, the required strength of the correcting lens (and the consequent magnification) will depend upon how old they are, as we have seen.

Bifocals are spectacles that generally contain lenses with two portions, one of a power appropriate for distance vision and another appropriate for near vision. It is normal for the majority of the lens to be appropriate for distance vision, with a small portion at the bottom of each lens of a more positive power needed for near vision, as depicted in Figure 4.4. The reason for this is that when reading it is natural to look down, and thus to look through the lower part of the spectacle lens. Other near tasks performed with the hands, such as threading a needle, are usually treated in the same way — the head is lowered and the eyes are lowered further still.

A relatively recent change to spectacle technology has been the development of multi-focal lenses. These are like bifocals in that they have a distance portion and a near portion, but they differ in that a smooth modification occurs in power from one to the other. Multifocal lenses do not have the sharp transition of bifocals and are generally considered cosmetically superior in this respect. In addition, the smooth change in

---

* This can be seen by looking at a straight line through an off-center portion of a spectacle lens. The magnification, or minification, will make the line appear in a different place.

Single Vision

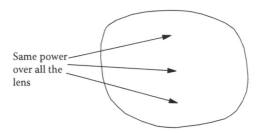

Bifocals

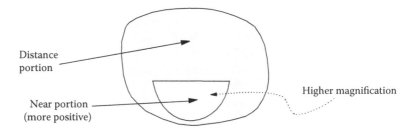

Multifocal lens

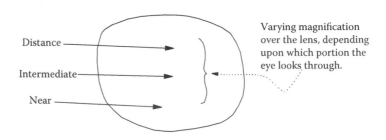

Figure 4.4 Different types of spectacle lenses.

power means that intermediate distances will be in focus at some part of the lens.

4.3.3 Problems with the Wearing of Spectacles

4.3.3.1 Single Vision

The problems encountered with distance and near spectacles, in the context of falls, differ. Problems with distance spectacles are primarily encountered on first wearing, or when a prescription is altered. In both of these cases, the change in magnification produces a distortion of the visual world (as well as a change in sharpness), which can be disorientating. Furthermore, the prescription in the two eyes may not be the same and, consequently, the distortion seen by each will differ, compounding the problem. The difficulty is particularly apparent when the head is moved, as such a movement will produce a different movement of the visual world from that seen previously and from that expected. In addition, hand–eye coordination will be affected adversely, as objects are not always exactly where they appear to be. Clinical experience tells us that, for most people, adaptation to the new appearance of the world usually occurs within a day or so, although for some people it takes longer and (rarely) some people are never able to adapt.

The problems encountered with near vision spectacles are less complex, being associated primarily with blur, instead of with spatial distortion. A near prescription is generally of relatively low positive power and is usually similar in the two eyes, giving the same overall magnification in each. Thus, when wearing spectacles designed for close work, it is the inability to see distant objects clearly, which is more of a problem than spatial distortion, particularly as bodily movements will tend to be small when a near prescription is worn.

4.3.3.2 Bifocals and Multifocals

Anecdotal evidence supports the suggestion that bifocal (and multifocal) lens wearers might be at increased risk of falls in circumstances such as using stairs. The underlying ideas behind this hypothesis are straightforward. First, localisation can be difficult. If a person looks down to see the edge of a step, he or she will do so by looking through the reading portion of the lenses, and thus the stair edge will be blurred. Because it is more difficult to localise a blurred edge than a clear one, the risk of "missing" the edge increases. In addition, the increased magnification of the near portion, in comparison with the distance portion, will affect localisation adversely. Second, detection can be impaired. In addition to the stair edge being blurred, any object on the stairs viewed through the "near" portion of the lens will also be blurred. The detection of the object's outline is important for its conspicuity (Section 4.2), and as blurring will reduce the appearance of the outline, it will consequently increase the likelihood of it not being noticed.

A recent study by Davies et al. (2001) provides supporting evidence that the wearing of bifocal or multifocal lenses can be a contributory factor in falls. The study examined hospital patient interviews using the Merseyside Accident Information Model and found a significant association between accidents in which the first event was the missed edge of a step and the wearing of bifocals or multifocal lenses. The evidence discussed by Davies et al. (2001) indicated that stepping down (as opposed to stepping up) increased the risk of "missed edge" accidents.

In "stepping down" accidents, the problems of bifocal wear and multifocal wear are somewhat different. For both, the wearer will see a clear image of distant objects in the world through the larger upper portion of the lenses. For the bifocal wearer, looking down at a step through the lower segment of the lens may involve looking at a blurred image, as the range of clear vision may not include the step. In addition, a disjunction may occur at the position where the lens power changes. Because the two portions produce slightly different size images, there cannot be continuity between the two images seen through the different lens portions. For the multifocal lens wearer, the same issue of blur could arise, but no sharp divide exists between the distance and near portions of the lenses. Instead, a transition zone exists where the power changes from one to the other. The positive advantage gained from this lens form is that with appropriate positioning of the head, a portion of the lens can be found that will provide a clear image of the step. The cost, however, is that the person may experience image distortion, which is particularly noticeable when the head is moved because the image movement is not consistent over the whole of the lens. The resulting distortion is often termed "swim."

4.3.3.3 The Wearing of the Wrong Spectacles (or None at All)

Optically, the power of a lens is the reciprocal of its focal length, in metres, and is measured in a unit called a Dioptre (D). A lens that focusses light from a far-off object at a distance of 1 m would be a 1 D lens. Optical power is additive when it is measured in this way, and so two such lenses put together would produce a 2 D lens, which would focus this light at a distance of 0.5 m.

Few people watch television at a distance closer than 3 m. Surprisingly, 3 m is little different optically from the distance to the sun; the optical distance of the former is 1/3 D while that of the latter is 0 D, a difference of only 1/3 D. On the other hand, if a person's hands are 1/3 m from the eye and his or her feet are 1.5 m away, then the difference between them optically is 2 1/3 D (3 D–2/3D). A pair of near spectacles will blur distance vision in the same way that, without accommodation, near vision will be blurred when distance spectacles are worn. For looking at objects at ground level, however, the distance pair will produce an appreciably less-blurred image than the near pair. Therefore, it would be expected that the wearing of near vision spectacles while moving around an indoor environment would be more hazardous than the wearing of a distance correction.

Evidence exists that among older people, the wearing of incorrect spectacles is common. Brace et al. (2003) found that among 177 older people interviewed, 8% admitted to sometimes wearing their reading glasses when walking around the house or wearing their spouse's glasses if they could not find their own. Some people refer to their "indoor" and "outdoor" spectacles, where the former are actually for near vision and the latter for distance vision, suggesting that an inappropriate correction might be worn inside the house. In addition, it is all too easy to misplace a pair of spectacles and to subsequently try to manage without them.

4.3.4 Loss of Visual Ability through Disease

People with a visual impairment generally find walking in unfamiliar areas difficult, and consequently will often limit their independent travel (Turano et al. 1999). Even people with moderate visual impairment walk slower and have more bumps and stumbles than those with normal vision (Lovie-Kitchen et al. 1990). Conditions such as retinitis pigmentosa (RP) will affect vision in characteristic ways, in this case by restricting the visual field and producing "tunnel vision." These changes then bring about changes in behaviour. Turano et al. (2001), for example, reported that people with RP who walked an unfamiliar obstacle-free route fixated over a larger area of the visual field than did normal participants. Those persons with normal vision generally directed their gaze ahead at their goal, whereas those with RP directed their gaze at objects on the walls, downward, or at the overall layout of the environment.

Other diseases will similarly produce characteristic visual losses, and changes in behaviour. Age-related maculopathy, for example, affects a person's central vision: They can see better when they look to the side of an object than when they look straight at it. People will generally produce coping strategies to overcome the reduction in their ability, but the effect of these cannot be generalised because they differ depending on the circumstances.

4.4 Static and Dynamic Situations

The importance of vision in maintaining static balance was demonstrated by Lee and his colleagues in the 1970s (e.g., Lishman and Lee 1973; Lee and Aronson 1974; Lee and Lishman 1975) using a "swinging room." Such a room consists of a bottomless box, suspended from the ceiling, which can be swung backward and forward. The participant stands on the floor, surrounded by normal-looking wallpapered walls, unaware that the "room" surrounding them can move. A movement of the (whole) room produces the same visual changes (optic flow pattern) that would occur if the person were actually swaying. These changes then, in turn, produce an opposite response from the participant who is unwittingly trying to stop swaying.

The contribution of vision to producing a feeling of motion is apparent in the situation many of us have experienced of sitting in a stationary train in a station, while another train is slowly moving alongside. During normal motion of the train, we do not experience acceleration, which is one of the aspects of movement our vestibular system detects, and, if it were not for the bumps, we might otherwise imagine ourselves to be stationary. The sensation of motion arises primarily from the visual information of the world passing the train window, so that when the world outside the train window (i.e., the other train) is seen to move, we have the sensation of (opposite) movement ourselves.

This sense of bodily movement develops early in life. Lee and Aronson (1974) placed toddlers (age 13–16 months) within the swinging room, and observed responses of backward movement when the wall swung toward them and forward movement when it moved away from them. These movements were compensatory — a backward movement of the wall would make the child feel that they were moving forward and the compensatory movement would then be backward, indicating that the link between motion and the consequent visual changes is already established at this young age.

As we move through the environment, the world moves past us in a consistent manner. The image on the retina expands as we move forward, producing what Gibson termed "Optic Flow" (see Gibson 1968). Anyone familiar with the opening sequence of the television show "Star Trek" (or anyone with "Starfield Simulation" as their Microsoft Windows™ screen saver) will have seen an example of optic flow. In these examples, it is the movement of the stars that provides the feeling of movement toward the centre of the screen.

This feeling of motion that comes about from changing visual information is termed "vection." One of the interesting aspects of vection is the powerful role that the peripheral visual field plays. For geometrical reasons, optical flow changes in the periphery are larger than those of central vision. Although central vision is important for perception, in that we turn our eyes to look directly at an object when we want to see its details, peripheral vision is important in providing visual information to be compared with the information from the vestibular system. As we move, we may direct our eyes and focus our attention on an object in front of us to see its details and its location, but as we do, so it is our peripheral vision that provides us with the visual information about our movement. Thus, restricting peripheral vision is likely to affect balance negatively (Wade et al. 2004).

The principal clues provided by vection are the relative velocity and the direction of movement of objects in our visual field. As we move forward toward an object on which we are fixating, those objects above eye height move upward in our visual field, and those below eye height move downward. Similarly, those to the left and right flow to the side in opposite directions and move further away from the midline.

Most of us know from observing the outside world when sitting in a moving vehicle that nearer objects move past the window faster than further objects. This is termed "motion parallax." When moving through an environment, considerable information on motion parallax is collected by our visual sense (in conjunction with our other senses), and the consistency of this is one of the factors that allow us to monitor our own movement. Motion parallax, however, also provides information about the world around us, particularly with respect to where objects are positioned relative to each other.

The importance of this information can be seen from the results of a recent Dutch investigation. Alferdinck et al. (2001) studied staircase design because of concerns that the intricate pattern of a large stairway in the city of Vlaardingen could camouflage the stair edges. The same issue has also been raised by Hill et al. (2000) in the context of stairs in the home environment (see Figure 4.5 taken from this study).

In the Dutch investigation, two stair coverings were compared, one patterned and one patternless. Walking time, number of steps taken, and number of missteps were all measured and, surprisingly, there was no difference between the two conditions. The authors noted that from the viewpoint of someone who is descending the stairs, the transitions between steps are visible because of the relative movement of the successive steps with respect to each other (motion parallax).

The subjects used in Alferdinck et al.'s (2001) study were all of working age (16 were under 48 years old, and the other two were 52 and 60 years old); all subjects knew they were participating in an experiment. Consequently, it is necessary to be circumspect in transferring these results to a nonexperimental context and to an older population. Instead of indicating that no problem exists with patterned staircases, from Figure 4.5,

Figure 4.5   Step edges on stairs camouflaged by carpet patterning. (From Hill, L.D., Haslam, R.A., Howarth, P.A., Brooke-Wavell, K., and Sloane, J. (2000) *Safety of older people on stairs: behavioural factors.* **Department of Trade and Industry; London. DTI ref. 00/788.)**

it is clear that potential problems exist. The results suggest how people can overcome these problems — they pay attention and they use motion parallax to provide change-of-depth information.

Here then is another difficulty for the spectacle wearer. Any lens will distort the world, by magnifying or minifying it, and this distortion increases as one moves away from the centre of the lens. Consequently, the optic flow information available from motion parallax is different when wearing spectacles from when they are not worn, and it has been suggested that this is the reason why some people initially experience giddiness and nausea when they change their spectacle prescription.

Potentially, the situation is worse for bifocal, trifocal, or multifocal wearers because the motion parallax information will be different in different sections of the lens. The multifocal lens wearer in particular could suffer because a continuous change in power, and thus a continuous change in magnification, occurs over the lens. Consequently, any head movement, altering the portion of the lens that the person is looking through, will be accompanied by a magnification change. Although these changes will be relatively small, they are still large enough to ensure that some people feel unable to wear multifocal lenses.

4.5 Behaviour and the Environment

The issue of conspicuity was raised in Section 4.2, where it was noted that detection of an object depends not only upon the characteristics of the object but also on the context in which it is seen. Thus, the role of lighting is not simply to illuminate the environment, revealing its form, but also to reveal the presence, location, and shape of objects within that environment. We need to go further here than simply considering the amount of light, important though this is, and consider the quality of the lighting. The spectral aspects of light can change the appearance of objects and, importantly, the directionality of the light reveals shape and form.

Compare the two photographs in Figure 4.6, which are of the same object seen in two different lighting conditions. In one the lighting is "flat," the scene is shadowless, and the object appears to lack shape. In the other, the object appears to have form and solidity, and the presence of the shadows provides both an enhanced outline to the object and gives it a three-dimensional appearance.

In general, the higher the overall light level the better people can see, unless glare is apparent. At higher light levels, the ability to see detail and to judge depth improves, and therefore it is reasonable to suppose the risk of falling will decrease. Conversely, the lower the light level the

Figure 4.6 Object viewed under flat (top) and directional (bottom) lighting conditions.

greater the risk of falling. To date, little empirical evidence exists about this relationship, but it is reasonable to suppose that in normal, daytime ("photopic") light levels the change in risk with illumination level is small. As one approaches dusk-like conditions ("mesopic"), visual performance decreases dramatically, and it appears likely that there will be a commensurate decrease in ability to monitor the walking environment. This is certainly the case with driving, where increases in road lighting appear to progressively reduce nighttime accident rates until a road surface luminance level of about 1 cd/m^2 is reached, after which little further improvement occurs (Barbur et al. 1998; Plainis et al. 1997). Therefore, the primary concern at higher light levels will be how well light reveals the form of objects, whereas at lower light levels, it is simply the revealing of their presence that is important.

In addition to the role played by the provision of light, it is also necessary to consider how people behave in terms of their use of it. Although most people would accept readily that they see better indoors at night when the light has been switched on (and so are less likely to have a fall), Haslam et al. (2001) reported how older people described getting up at night to use the toilet without turning on the light to avoid disturbing a sleeping partner. After being asleep, our eyes are usually dark-adapted and, if the environment is familiar, many people believe they have enough vision to accomplish their task without a problem; however, the risks involved in this behaviour are obvious.

4.6 Interaction between the Visual and Vestibular Systems

The role of vision in providing information about movement has been described previously (Section 4.4). Inputs from our other senses allow us to interpret visual information correctly, identifying whether it is ourself that is moving or whether it is objects in the environment. For example, movement of the head and eyes, movement of the eyes alone, and movement of the environment could all produce identical movements of the image on the retina (the light-sensitive layer at the back of the eye). The monitoring of signals provided by the vestibular and somatosensory systems, as well as the command signals to the extraocular muscles that move the eyes, allows the brain to determine which of these has actually occurred. The strong link between the visual and vestibular systems can be seen in the vestibulo-ocular reflex, where movement of the head in one direction is accompanied by an equal but opposite movement of the eyes in the opposite direction, maintaining fixation.

Conditions where this consistency can fail do exist, however, leading to a situation where the senses give us conflicting information. The introduction of a new spectacle prescription is one such instance (as discussed in Section 4.3.3), in which the altered magnification will make objects appear to be in a slightly different place from where they are expected to be. In these circumstances, the visual and vestibular systems need time to recalibrate, which they can usually do successfully even if the prescription in the two eyes differs (Lemij and Collewijn 1991). Another such situation occurs when people wear headsets to play computer games that provide vection. Here, the visual system reports bodily movement but the vestibular and somatosensory systems both report that the body is stationary. Here, the "sensory conflict" (Reason and Brand 1975) results in feelings of disorientation followed by feelings of nausea (e.g., Howarth and Finch 1999).

Clearly, an incorrect interpretation of either bodily motion or movement of the environment can lead to a fall. The relative importance of the visual, vestibular and somatosensory signals in determining the overall feeling of motion (and thus interpreting the information correctly) appears to vary between people. For most individuals the visual information appears to play a vital role, as indicated at the beginning of this chapter. Interestingly, for some people, normal visual information can cause disorientation at times, and this phenomenon has been termed "visual vertigo" (see Ciuffreda 1999; Guerraz et al. 2001). For these individuals, disequilibrium attacks can be brought about by normal visual stimuli such as repetitive wallpaper, traffic flow, or passing food displays in supermarkets. For some people repeated patterns, such as those found on escalator steps, can lead to feelings of instability (Cohn and Lasley 1985) as can changing lighting conditions (Simoneau et al. 1999). Guerraz et al. (2001) concluded that in such patients, visual vertigo arises because of "enhanced visual dependence," an over-dependence on visual information that reflects an "idiosyncratic perceptual style" in which the individual is overly dependent on vision for perception and postural control. This "physiological visual vertigo," so named because no evidence of pathological dysfunction exists in any of the sensory systems, is poorly understood by clinicians, researchers, and patients alike. It follows that its prevalence is likely to be under-reported, and consequently little is known about its role, and importance, in falls.

4.7 Summary

The role of vision in falls is most obvious when it fails: Unseen objects or obstacles become hazards to be slipped on or tripped over. Environmental changes, such as increasing levels of light and altering its directionality, can help in this respect, as can ensuring that lamps are switched on when needed and that correct spectacles are worn.

Less obvious, but no less important, is the primary role vision has in providing subconscious information about the body's position in space. The experiments of Howard and his co-workers (e.g., Groen et al. 1999) have demonstrated the importance of the visual perception of the vertical (see Howard 1986a, 1986b). Many of their participants, strapped into a chair in an artificial moving room, were completely unaware of having been rotated, because the visual stimulus remained unchanged. Here, the visual information overrides the vestibular and somatosensory inputs, demonstrating the importance of vision in providing our model of the world and where we are in relation to our environment. In the absence of accurate information about this, it will not be long before our pride comes before a fall.

References

Alferdinck, J.W.A.M., Walraven, J., and Kooi, F.L. (2001) Risk evaluation of a decorated staircase. In *TNO Human Factors Research Institute internal report*. TNO Soesterberg, The Netherlands.

Barbur, J.L., Harlow, A.J., Smith, P., and Hurden, A. (1998) Visual performance in the mesopic range. In *Non-invasive assessment of the visual system* (Technical Digest Series), Optical Society of America, Washington D.C., 1, 140–143.

Brace, C.L., Haslam, R.A., Brooke-Wavell, K., and Howarth, P.A. (2003) *The contribution of behaviour to falls among older people in and around the home*. Loughborough University for Department of Trade and Industry, Leicestershire (http://www.lboro.ac.uk/departments/hu/groups/hseu/publications).

Brooke-Wavell, K., Perrett, L.K., Howarth, P.A., and Haslam, R.A. (2002) Influence of the visual environment on the postural stability in healthy older women. *Gerontology*, 48: 293–297.

Ciuffreda, K.J. (1999) Visual vertigo syndrome: clinical demonstration and diagnostic tool. *Clinical Eye and Vision Care*, 1: 41–42.

Cohn, T.E. and Lasley, D.J. (1985) Visual depth illusion and falls in the elderly. *Clinics in Geriatric Medicine*, 1: 601–620.

Davies, J.C., Kemp, G.J., Stevens, G., Frostick, S.P., and Manning, D.P. (2001) Bifocal/varifocal spectacles, lighting and missed-step accidents. *Safety Science*, 38: 211–226.

Gibson, J.J. (1968) What gives rise to the perception of motion? *Psychological Review*, 75: 335–346.

Groen, E.L., Howard, I.P., and Cheung, B.S.K. (1999) Influence of body roll on visually induced sensations of self-tilt and rotation. *Perception*, 28: 287–297.

Guerraz, M., Yardley, L., Bertholon, P., Pollak, L., Rudge, P., Gretsy, M.A., and Bronstein, A.M. (2001) Visual vertigo: symptom assessment, spatial orientation and postural control. *Brain*, 124: 1646–1656.

Haslam, R.A., Sloane, J., Hill, L.D., Brooke-Wavell, K., and Howarth, P.A. (2001) What do older people know about safety on stairs? *Ageing and Society*, 21: 759–776.

Hill, L.D., Haslam, R.A., Howarth, P.A., Brooke-Wavell, K., and Sloane, J. (2000) *Safety of older people on stairs: behavioural factors.* Department of Trade and Industry, London. DTI ref. 00/788.

Howard, I.P. (1986a) The perception of posture, self motion, and the visual vertical. In K.R. Boff, L. Kaufman, and J.P. Thomas (eds.). *Handbook of perception and human performance: vol. 1 sensory processes and perception.* John Wiley & Sons, New York.

Howard, I.P. (1986b) The vestibular system. In K.R. Boff, L. Kaufman, and J.P. Thomas (eds.). *Handbook of perception and human performance: vol. 1 sensory processes and perception.* John Wiley & Sons, New York.

Howarth, P.A. and Finch, M. (1999) The nauseogenicity of two methods of navigating within a virtual environment. *Applied Ergonomics*, 30: 39–45.

Lee, D.N. and Aronson, E. (1974) Visual proprioceptive control of standing in infants. *Perception and Psychophysics*, 15: 529–532.

Lee, D.N. and Lishman, J.R. (1975) Visual proprioceptive control of stance. *Journal of Human Movement Studies*, 1: 87–95.

Lemij, H.G. and Collewijn, H. (1991) Long-term nonconjugate adaptation of human saccades to anisometropic spectacles. *Vision Research*, 31: 1939–1954.

Lishman, J.R. and Lee, D.N. (1973) The anatomy of visual kinaesthesis. *Perception*, 2: 287–294.

Lord, S.R. and Dayhew, J. (2001) Visual risk factors for falls in older people. *Journal of the American Geriatrics Society*, 49: 508–515.

Lovie-Kitchen, J., Mainstone, J., Robinson, J., and Brown, B. (1990) What areas of the visual field are important for mobility in low vision patents? *Clinical Vision Science*, 5: 249–264.

Plainis, S., Chauhan, K., Murray, I.J., and Chatman, W.N. (1997) *Retinal adaptation under night-time driving conditions.* Vision in Vehicles VII, Marseilles, France, 13–17 September 1997.

Reason, J.T. and Brand, T. (1975) *Motion sickness.* Academic Press, London.

Simoneau, M., Teasdale, N., Bourdin, C., Bard, C., Fleury, M., and Nougier, V. (1999) Aging and postural control: postural perturbations caused by changing the visual anchor. *Journal of the American Geriatrics Society*, 47: 235–240.

Turano, K.A., Geruschat, D.R., and Quigley, H.A. (1999) Getting around with vision impairment. In *15th biennial eye research seminar book*, Research to Prevent Blindness, New York.

Turano, K.A., Geruschat, D.R., Baker, F.H., Stahl, J.W., and Shapiro, M.D. (2001) Direction of gaze while walking a simple route: persons with normal vision and persons with retinitis pigmentosa. *Optometry and Vision Science*, 78: 667–675.

Wade, L.R., Weimar, W.H., and Davis, J. (2004) Effect of personal protective eyewear on postural stability. *Ergonomics*, 47: 1614–1623.

Chapter 5

Ageing and Falls

Stephen Lord, Catherine Sherrington,
and Hylton Menz

CONTENTS

5.1 Introduction

Susceptibility to falling increases considerably with ageing (Lord et al. 2001). One in three older people living in the community are likely to fall one or more times in a year, and falling rates are higher in older people living in residential care. The actual definition of a fall in older people, however, has been open to some debate. A frequently used definition is "unintentionally coming to the ground or some lower level and other than as a consequence of sustaining a violent blow, loss of consciousness, sudden onset of paralysis as in stroke or an epileptic seizure" (Gibson et al. 1987). This definition excludes overwhelming external disturbances that result in an older person being knocked over and major internal disturbances that cause an older person to collapse instead of fall. Depending on the focus of study, however, some researchers have used a broader definition of falls to include those that occur because of dizziness and syncope. Although falls are often referred to as accidents, it has been demonstrated statistically that they are not random events (Grimley-Evans 1990). This implies that causal processes are involved, and many studies have been undertaken with the aims of identifying fall risk factors. In broad terms, these risk factors can be classified as being of a demographic, psychosocial, medical, physical or environmental nature (Lord et al. 2001), and strategies and guidelines for addressing these factors to prevent falls have recently been devised (Feder et al. 2000; American Geriatrics Society et al. 2001; Lord et al. 2001). This chapter reviews the work that addresses the epidemiology of falls in older people, physical and environmental fall risk factors, and uniform and multifaceted strategies for preventing falls in this group.

5.2 Epidemiology of Falls in Older People

At least one in three adults age 65 and older will fall one or more times each year. For example, in the Randwick Falls and Fractures Study conducted in Australia, we found that 39% of 341 community-dwelling women reported one or more falls in a 1-year follow-up period (Lord et al. 1993). Similarly, in a study of 761 subjects age 70 and over undertaken in New Zealand, Campbell et al. (1989) found that 40% of the women and 28% of the men fell at least once in the study period of one year, an overall incidence rate of 35%. Falling rates in residential aged care facilities are

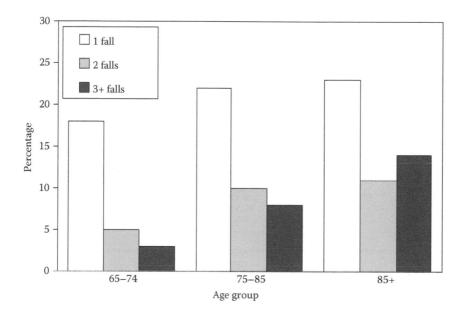

Figure 5.1   Falls rates according to age. (Adapted from Lord, S.R., Ward, J.A., Williams, P., and Anstey, K.J. (1993). *Australian Journal of Public Health* **17(3): 240–5.)**

even higher, with 50–60% of residents falling at least once each year (Robbins et al. 1989; Yip and Cumming 1994; Liu et al. 1995).

Falling rates increase beyond the age of 65 years. Figure 5.1 depicts the proportion of women in different age groups who reported falling once, twice, or three times during the 12-month Randwick Falls and Fractures Study (Lord et al. 1993). Similar incidence rates have been reported in the United Kingdom (Prudham and Evans 1981), United States (Nevitt et al. 1989), and Europe (Luukinen et al. 1994).

Injury is a potentially serious outcome of falling in older people. One-quarter to one-half of all falls among community dwellers cause some injury, 10–15% of falls are associated with serious injury, 2–6% with fractures and around 1% with hip fractures. The most common self-reported injuries include superficial cuts and abrasions, bruises, and sprains. The most common injuries that require hospitalisation are femoral neck fractures, other fractures of the leg, fractures of radius, ulna, and other bones in the arm and fractures of the neck and trunk (Gibson et al. 1987; Lord 1990; Speechley and Tinetti 1991).

The proportion of people suffering fall-related injuries increases with age (Prudham and Evans 1981; Nevitt et al. 1989; Luukinen et al. 1994). In the Randwick study (Lord et al. 1993), 23% of women age 65 to 74 suffered

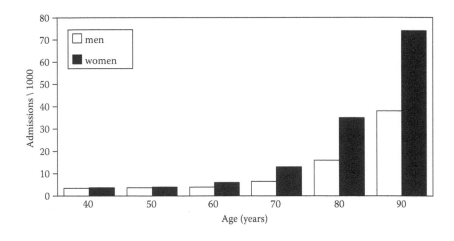

Figure 5.2   Hospital admissions for falls according to age and gender. (Adapted from Lord, S.R. (1990). *Medical Journal of Australia* **153(2): 117–8.)**

one or more injuries, and this number was 35% for women age 85 and over. Interestingly, "vigorous" older people suffer injuries that are more serious when they fall than frailer subjects (Speechley and Tinetti 1991). It is suggested that this is because people who are more vigorous fall while undertaking activities that require more displacement of the centre of gravity.

Falls are the leading cause of injury-related hospitalisation in persons age 65 and over. Hospital admissions resulting from falls are uncommon in young adulthood, but with advancing age, the incidence of fall-related admissions increases at an exponential rate. As presented in Figure 5.2, beyond 40 years of age the admission rate due to falls increases consistently by 4.5% per year for men (doubling every 15.7 years) and by 7.9% per year for women (doubling every 9.1 years) (Lord 1990). In those age 85 and over, the levels have reached 4% per annum in men and 7% per annum in women.

Several studies have reported that a history of falls increases the likelihood of an older person moving to a long-term residential care institution (Tinetti and Williams 1997; Donald and Bulpitt 1999). For example, one study found that after controlling for other factors, the odds of entry into care were three times higher for those who had suffered an injurious fall than those who had not (Wilkins 1999).

5.2.1 Circumstances of Falls among Older People

Around half of all falls experienced by healthy community dwellers occur within the person's own home (Nevitt et al. 1989; Campbell et al. 1990;

Lord et al. 1993; Luukinen et al. 1994). Campbell et al. (1990) found that 16% of falls occurred in the person's own garden, 21% in a bedroom, 19% in the kitchen, and 27% in the lounge/dining room. Our group found that 6% of falls occurred while using the shower or bath, 3% off a chair or ladder, 6% on stairs, and 26% while walking on a level surface (Lord et al. 1993).

The location of falls varies according to age and gender. With increasing age, more falls occur inside the home on level surfaces (Lord et al. 1993). Women are more likely to fall within their usual residence, whereas men are more likely to fall in their own garden (Campbell et al. 1990). Similarly, men are more likely to suffer a hip fracture outdoors (Allander et al. 1996). Between 21% (Campbell et al. 1990) and 53% (Lord et al. 1993) of falls are attributed to trips and slips. Although up to 14% of people are unable to identify the cause of their falls, 21% report losing their balance and 6% report their legs giving way (Lord et al. 1993).

Several authors have also investigated what the person was doing when they fell. Tinetti et al. (1988) classified falls according to the amount of displacement of the subject's centre of gravity involved in the activity associated with a fall. They report that 57% of falls occurred during activities that mildly displaced the person's centre of gravity (e.g., standing still, basic activities of daily living, walking); 34% of falls occurred during moderately displacing activities (e.g., getting up or sitting down, reaching up or down, stepping up or down); and 5% of falls occurred during activities involving marked displacement of the centre of gravity (sports and climbing on chairs, ladders or other objects). Most falls among community dwellers occur during "periods of maximum activity" in the morning and afternoon (Luukinen et al. 1994; Berg et al. 1997) with only about 20% of falls occurring between 9 p.m. and 7 a.m. (Campbell et al. 1990).

5.3 Physical Risk Factors for Falls

The maintenance of postural stability is a complex ability that relies on the interaction of sensory, motor, and integrative systems (see Chapter 2). Consequently, impaired vision, sensation, strength, and reaction time may all contribute to poor balance. A growing body of evidence indicates that normal ageing is associated with impairments in each of these systems, and that this impaired function manifests as an impaired ability to maintain balance when performing a range of functional tasks. Generally, the more challenging a task is to the postural control system, the greater the effect of age and the stronger the relationship to falls. The following section discusses the relationship between ageing and impaired balance ability, and summarises the evidence for the association between sensori-motor impairment and risk of falling.

5.3.1 Ageing, Impaired Balance, and Falls

5.3.1.1 Standing

Normal standing is characterised by small amounts of postural sway, the control of which requires an integrated reflex response to visual, somatosensory, and vestibular inputs (Fitzpatrick et al. 1994). Sway reportedly increases significantly with age (Sheldon 1963; Hasselkus and Shambes 1975; Lord and Ward 1994) and has been demonstrated as a useful predictor of falls in older people (Lord et al. 1991a; Lord et al. 1994b). In our studies, we have found that older people who fall exhibit greater sway both with their eyes open and closed. The differences between fallers and non-fallers, however, is particularly evident when subjects stand on a medium-density foam rubber mat, which decreases ankle support and alters proprioceptive feedback from the foot and ankle (Lord et al. 1991a; Lord et al. 1994b). Other standing tests reveal similar differences between fallers and non-fallers. Near tandem standing (standing with one foot in front of the other) (Lord et al. 1999), unipedal standing (standing on one foot) (Vellas et al. 1997), and forward leaning ability (Duncan et al. 1992) have all been demonstrated to discriminate between fallers and non-fallers.

5.3.1.2 Responses to Perturbation

In recognition that some falls occur in response to the application of an external force, a number of researchers have investigated postural responses to mechanically applied perturbation in older people, using translating platforms. Compared with younger people, older people generally have more difficulty generating efficient postural responses to unexpected perturbations, often requiring multiple corrective steps to reestablish postural stability (Maki et al. 1990; McIlroy and Maki 1996). These techniques have provided considerable insight into the mechanisms underlying reflex responses to unexpected threats to stability, however, their ability to predict falls is limited. Two prospective studies have found that measurements of sway during unperturbed standing are more able to discriminate between fallers and non-fallers than responses to perturbation (Maki et al. 1990; Maki et al. 1994).

5.3.1.3 Gait

The maintenance of equilibrium when walking is a particularly challenging task for the human postural control system for two main reasons:

1. Two-thirds of the body's mass is situated two-thirds of body height above the ground.
2. For a major period of the walking cycle, the body is supported by a single limb with the centre of gravity passing outside the base of support (Winter 1995).

Subsequently, the potential for loss of balance when walking is considerable, and it is therefore not surprising that up to 75% of falls occur during ambulation (Cali and Kiel 1995; Norton et al. 1997).

. Numerous studies have revealed that older people have significantly different gait patterns to younger people, including reduced velocity and cadence (Oberg et al. 1993; Lord et al. 1996; Bohannon 1997), increased cadence variability (Lord et al. 1996; Hausdorff et al. 1997), a larger out-toeing foot placement angle (Murray et al. 1969), decreased hip range of motion (Crowinshield et al. 1978), and decreased ability to generate ankle power for forward momentum (Kerrigan et al. 1998). Some of these gait alterations, particularly decreased velocity and increased variability of cadence, have been found to discriminate between fallers and non-fallers in prospective investigations (Lord et al. 1996; Hausdorff et al. 1997).

In response to the observation that a large proportion of falls are related to trips and slips, a number of studies have been performed to evaluate the ability of older people to navigate obstacles and slippery surfaces when walking. Compared with younger subjects, older people require more time to successfully step over obstacles when walking (Chen et al. 1991), clear the ground by a smaller margin when stepping onto a raised surface (Begg and Sparrow 2000), take longer to change direction in response to a visual stimulus (Cao et al. 1997), and are less able to adjust their stride length to compensate for a heel strike slip when walking (Tang and Woollacott 1998). Clearly, the inadequacy of these postural responses places the older person at an increased risk of falling when walking under challenging conditions. Furthermore, evidence from recent studies indicates that the inadequacy of an older person's postural response may be further exacerbated when his or her attention is divided between multiple tasks. For example, older people are less able to successfully navigate obstacles when verbally responding to a visual stimulus (Chen et al. 1996), which suggests that with advancing age, the maintenance of stability becomes more cognitively dependent.

The available research evidence therefore suggests that older people have difficulty maintaining balance when performing various tasks, and subsequently are at greater risk of losing balance during daily activities. Given that balance ability is dependent on the interaction of various sensory and neuromuscular systems, impaired physiological function in each of these domains may increase the risk of falling.

5.3.2 Sensori-Motor Impairment and Risk of Falling

It has long been recognised that frail older people with multiple chronic illnesses have higher rates of falls than do active, healthy older people. Attributing a degree of falls risk to a specific medical diagnosis is problematic, however, because the relative severity of the condition may vary considerably between individuals. Furthermore, a decline in sensori-motor function due to age, inactivity, medication use, or minor pathology may be evident in older people with no documented medical illness. In response to this problem, many researchers have taken a physiological instead of a disease-oriented approach to evaluating falls risk factors. This approach involves direct assessment of sensori-motor capabilities instead of documenting the presence or absence of a diagnosed disease. For example, in an older person with cataracts and associated visual impairment, the identified risk factor is impaired vision (e.g., poor visual acuity or depth perception), instead of cataracts per se. In this context, the following section summarises the evidence for the association between sensori-motor impairments and risk of falling in older people.

5.3.2.1 Vision and Hearing

Visual input provides the brain with a continually updated reference frame regarding the position and motion of body segments in relation to each other (i.e., egocentric reference) and the environment (i.e., geocentric reference) (see Chapter 4). In the absence of accurate visual information, spatial awareness is impaired, leading to an increased likelihood of mis-judging obstacles such as steps or cracks in the footpath. Many researchers have found that visual abilities decline after the age of 40 years (Pitts 1982), and impaired vision is a risk factor for falls (Nevitt et al. 1989; Ivers et al. 1998) and hip fractures (Felson et al. 1989) in older people. In our studies, we have found that contrast sensitivity and depth perception are more important in predicting falls than visual acuity, as these visual abilities are more directly related to the ability to detect and discriminate obstacles in a cluttered environment (Lord et al. 1991b; Lord et al. 1994b; Lord and Dayhew 2001). Consistent with these findings, numerous studies have found that the presence of visual conditions, such as cataracts (Felson et al. 1989; Herndon et al. 1997) and glaucoma (Ivers et al. 1998), are associated with an increased risk of falling.

In contrast to the substantial evidence for the role of impaired vision in predisposing older people to fall, the evidence for impaired hearing increasing the risk of falls is equivocal. The large epidemiological studies that have measured multiple posited risk factors have not found poor hearing to be a risk factor (Tinetti et al. 1988; Campbell et al. 1989; Nevitt et al. 1991; O'Loughlin et al. 1993). Some studies have reported associations with impaired hearing and poor balance (Lichtenstein et al. 1988; Gerson et al. 1989) and dizziness (Tinetti et al. 2000) in older people; however, these studies did not include assessments of reduced vestibular functioning — a sensory impairment that could account for such associations because the vestibular system has a major role in balance control.

5.3.2.2 Peripheral Sensation

Sensory input from the extremities provides the central nervous system with important information regarding the position and movement of the limbs. Advancing age is associated with decreased vibration (Rosenberg and Adams 1958; Kenshalo 1986), tactile (Lord and Ward 1994; Stevens and Choo 1996) and joint position (Skinner et al. 1984; Gilsing et al. 1995) sense, and subsequently the ability to detect and control the movement of the legs when standing and walking is impaired. Older people with diabetic peripheral neuropathy exhibit even more pronounced sensory deficits (VandenBosch et al. 1995; Simoneau et al. 1996) and are more likely to fall than those with normal sensation (Richardson et al. 1992; Richardson and Hurvitz 1995); however, even in older people with no known neuropathic disease, we have found measures of peripheral sensation to be strong predictors of falls (Lord et al. 1991a; Lord et al. 1994a,b).

5.3.2.3 Vestibular Sense

The vestibular system detects the position and motion of the head in space and initiates corrective movements via vestibulo-ocular and vestibulo-spinal pathways. Research has revealed histological changes in the vestibular apparatus with advancing age (Rosenhall 1973) as well as subsequent reduced reactivity to caloric and rotational stimulation (Karlsen et al. 1981; Ghosh 1985); however, no studies have found tests of vestibular function to be strong predictors of falls in older people. This may be due to the insensitivity of the tests used for this purpose. For example, the commonly utilised Fukuda stepping test (Fukuda 1959), which involves measuring the displacement of a subject when he or she walks in place with his or her eyes closed, has not been found to be a predictor of falls (Lord et al. 1994b); however, many frail older people are unable to complete the test. Therefore, a need exists for more feasible and sensitive tests of vestibular function to clarify the role of age-related vestibular impairment in falls.

5.3.2.4 Muscle Strength

Both isometric and dynamic muscle strength decrease with age, particularly after the age of 60 (Petrovsky et al. 1975; Murray et al. 1980). Decreased muscle strength is particularly pronounced in the lower limb, which has important ramifications for stability and falls in older people (Pearson et al. 1985). In large community-based prospective studies, we have found that reduced quadriceps strength increases the risk of falls (Lord et al. 1994a,b) and fractures (Lord et al. 1992). Similarly, in residential aged care facilities, quadriceps and ankle dorsiflexor weakness have been found to be major risk factors for falling (Lord et al. 1991a; Luukinen et al. 1995).

5.3.2.5 Reaction Time

Reaction time is an important component of postural stability. To maintain balance under challenging conditions, it is necessary not only to detect the nature of the perturbation, but also react quickly to correct the imbalance (Grabiner and Jahnigen 1992). Reaction time has consistently been reported to decrease significantly with advancing age (Welford 1977), and in our studies, simple reaction time (i.e., pressing a switch in response to a visual stimulus) has been reported as an independent predictor of falls (Lord et al. 1991a). More recently, we have developed a novel measure of choice stepping reaction time that involves subjects stepping onto one of four panels when illuminated in a random order. Choice stepping reaction time performance was found to be strongly associated with a range of sensori-motor and balance variables, and was capable of discriminating between elderly fallers and non-fallers, suggesting that the test provides a composite measure of falls risk (Lord and Fitzpatrick 2001).

5.3.2.6 Relative Importance of Physiological Risk Factors

The preceding discussion indicates that impaired vision, peripheral sensation, strength, and reaction time are strong independent predictors of falls in older people. Falls are generally caused by a combination of multiple risk factors, however; and although a marked deficit in one of these systems may be sufficient to increase the chance of falling, a combination of mild impairments in multiple physiological domains may increase falls risk. Furthermore, it is likely that interactions occur between each of these systems, and that one system may be able to partially compensate for impairment in another. Nevertheless, considerable evidence indicates that the physiological systems that contribute to balance control decline with age, and impaired functioning of these systems is associated with falling.

5.4 Environmental Risk Factors

In addition to the physical abilities of the individual and the task being undertaken, environmental conditions are commonly suggested to be important in determining whether a person falls. Several prospective cohort (Tinetti et al. 1988; Campbell et al. 1989; Nevitt et al. 1989; Teno et al. 1990; Gill et al. 2000) and case control studies (Clemson et al. 1996; McLean and Lord 1996; Sattin et al. 1998) have now investigated the relative importance of environmental and other factors in falls. The majority of these studies have not found that the presence of environmental home hazards is a major risk factor for falls in older people; however, this does not mean that environmental factors are not commonly involved in falls when they occur. Around 45% of community-dwelling fallers report that their falls involved an environmental factor (Morfitt 1983; Tinetti et al. 1988; Nevitt et al. 1989). Similarly, in a survey of fall incident reports in a residential aged care facility, 50% of descriptions of falls involved an environmental factor (Fleming and Pendergast 1993).

Because approximately one-quarter to one-half of falls occur away from the person's own home (Campbell et al. 1990; Lord et al. 2001), environmental hazards within other people's homes and in public places may also be implicated in falls. The role of such hazards is difficult to assess in prospective risk factor studies. Some evidence indicates that environmental hazards play a greater role in falls that occur away from the home than within it. For example, one study found that 61% of falls occurring away from the home involved stairs or slipping and tripping hazards, whereas this number was 33% for falls at home (Nevitt et al. 1989). Temporary environmental hazards are also difficult to assess in prospective studies of environmental risk and have been identified as being important in interviews with people who have fallen (Connell and Wolf 1997). Temporary hazards may be more important than other types of environmental hazards as the person may not be able to anticipate and avoid them.

In addition, environmental factors can play a role in whether a fall will lead to serious injury. For example, a fall on stairs has been found to be associated with a twofold increase in risk of serious injury (Tinetti et al. 1995). Several studies have found that the risk of injury is greater for falls on hard surfaces (Nevitt et al. 1991; Nevitt and Cummings 1993).

The interaction between an environmental hazard and the person's physical abilities appears to play an important role in falls. Lawton (1980) describes a model of the interaction between an older person's compe-tence and the demands of the environment. A person must have a high competence level to cope effectively in an environment with high demands, whereas a person with a low competence level will be able to cope with an environment with low demands. A similar model can be used to describe the interaction between physical ability, environmental demand, and the resultant risk of falling (Lord et al. 2001). Those with high physical abilities can withstand a range of environmental challenges without falling, yet when faced with an extreme challenge (e.g., a patch of ice) they may still fall. Those with lower physical abilities can generally cope well in an environment that offers few challenges (e.g., by staying indoors), yet when these abilities are very poor, a fall will be experienced regardless of the safety of the environment (e.g., a fall while walking on a level surface in a bathroom with rails installed). For example, if a more able person trips on a cracked footpath, they may have the ability to recover from this and avoid a fall. Yet, a person with impairments in proprioception, reaction time, or muscle strength may not be able to recover in this way.

The type of environmental challenges that an older person chooses to expose themselves to, or in other words, the extent of a person's risk-taking behaviour, would be expected to be an important part of the interaction between the person and his or her environment. Indeed, a person's attitude to risk has also been found to be associated with falls (Studenski et al. 1994).

5.5 Interventions to Prevent Falls

In terms of interventions to prevent falls, research studies have followed two major approaches. The first approach has involved implementing interventions that have addressed single risk factors considered ame-nable to intervention (Lord et al. 2001). The main risk factor interven-tions implemented in such "uniform" study designs have been environmental modifications to promote safety in the home, exercise to promote strength and balance, as well as strategies for reducing hazardous medication use. The second approach has involved inter-ventions of a multifactorial nature, which target their prevention strat-egies to identified and often multiple risk factors. The final section of this chapter provides a brief overview of the randomised controlled trials that have examined the effectiveness of modifying single risk factors, and summarises the findings of the multifactorial interventions conducted to date.

5.5.1 Environmental Modifications

Environmental modification is often seen as an attractive falls prevention strategy. The homes of a majority of older people contain environmental hazards (Bray 1995; Carter et al. 1997), and many of these hazards are amenable to modification. Correction of these hazards is a one-off intervention that can be performed relatively cheaply (Smith and Widiatmoko 1998).

Two controlled clinical trials of home assessment and modification have been conducted (Liddle et al. 1996; Cumming et al. 1999), but only one of these trials (Cumming et al. 1999) reported falls as an outcome measure. The study by Cumming et al. was conducted among 530 community dwellers, most of who had been recently hospitalised. The intervention group received a home visit by an occupational therapist who assessed the home for environmental hazards and facilitated any necessary home modifications. Among those who had fallen in the year before the study, there was a significant reduction in the rate of falls in the intervention group compared with the control group. In fact, the relative risk reduction was 35%; however, the between-group difference among those who had not previously fallen was not significant. In addition, this study actually found a reduction in falls outside of the home. As discussed by the authors, this suggests that the home modifications may not have been the major factor in the reduction in falls rates. Other aspects of the occupational therapy intervention, which included advice on footwear and behaviour, may have played an important role.

The other, smaller, study also involved an assessment by an occupational therapist, but did not report such encouraging findings (Liddle et al. 1996). In this study of 167 older people, those found to require home modification or community services were randomised into either a group that had the occupational therapist's recommendations performed or a group that did not. At the 6-month follow-up, no differences were found on measures of health, mood, morale, life satisfaction or activities of daily living between the groups.

Although only limited evidence supports the role of home modification in falls prevention, more investigation is required. It may also be that home assessment and modification has a greater role to play among more disabled individuals (Campbell et al. 1989; Cumming et al. 1999). Indeed an appropriate environment may make the difference between someone being able or not able to complete a functional task, such as taking a shower. If a person has only marginal ability to complete such a task, the appropriate modifications may also greatly enhance his or her safety.

Modification of environmental hazards in public places is another potential falls prevention strategy. Common sense suggests that some falls could be prevented by the removal of obvious hazards, such as uneven pavements, slippery paths, and steep stairs, and by attempts to design and maintain safe indoor and outdoor environments for people with a range of physical abilities; however it is very difficult to rigorously evaluate the effectiveness of such strategies.

5.5.2 Exercise

For several years, it has been clear that exercise can modify key falls risk factors, such as decreased muscle strength, reduced speed, and poor balance in older people (Buchner et al. 1992; Lord et al. 1995), and good evidence now indicates that exercise can reduce the incidence of falls themselves (Gardner et al. 2000; Gillespie et al. 2003). Interestingly, effective interventions have involved disparate exercise regimes, including Tai Chi (Wolf et al. 1996), supervised strength and endurance training (Buchner et al. 1997), and physiotherapist- and nurse-directed home exercise (Campbell et al. 1997; Campbell et al. 1999a; Robertson et al. 2001). Further evidence that exercise is an effective intervention comes from a large, multifaceted intervention study that used home exercise as one of its components (Tinetti et al. 1994). These interventions resulted in significantly lower falling rates in the intervention subjects and improvements in intermediate measures, such as balance and transfer abilities, where exercise was likely to have been the agent of change.

Not all exercise interventions have been found to be effective in preventing falls, however, and a number of randomised trials have failed to demonstrate an effect of exercise on falls (Reinsch et al. 1992; MacRae et al. 1994; Mulrow et al. 1994; Lord et al. 1995). It is likely that some of the early interventions were not effective because the subjects were either at too high or too low a risk of falling for the interventions implemented.

Perhaps a better indication of the effectiveness of exercise as a falls prevention strategy comes from the meta-analysis of the seven FICSIT (Frailty and Injuries: Cooperative Studies of Intervention Techniques) trials that involved exercise (Province et al. 1995). Overall, the incidence of falls in these studies was reduced by about 10%. Among these trials, interventions classified as resistance, endurance, or flexibility training were not found to be effective in reducing fall rates, whereas exercises containing a balance component were. Overall, the interventions aimed at improving balance reduced falls by about 17%. Thus, it appears that exercise, particularly of a type that improves balance and enhances abilities to perform everyday functional tasks, is an important intervention strategy for maximising physical functioning and preventing falls in older at risk people.

5.5.3 *Modification of Medication Use*

Strategies to address medication use in older people include minimising the total number of drugs taken, assessing the risks and benefits of each drug, choosing drugs that are less centrally acting and do not produce postural hypotension, and reducing dose to the lowest possible effective level (Tinetti et al. 1994). Reducing medication use in older people is difficult, however, and may produce detrimental effects, particularly if performed too quickly. For example, rapid withdrawal of benzodiazepines may lead to confusion and restlessness (Bond and Schwartz 1984), which may impair an older person's ability to navigate obstacles in his or her environment (Campbell 1991). Nevertheless, the benefits of appropriate withdrawal of psychoactive medication can be significant, and are not necessarily associated with increased psychological problems (Avorn et al. 1992).

In a randomised controlled trial of gradual psychoactive medication withdrawal and home-based exercise, Campbell et al. (1999b) found a significant reduction in falls in the older community-dwelling women randomised to the medication withdrawal arms of the study. This is a very encouraging finding because the risk of falling for those who completed the trial was reduced by 65%. Considerable problems were encountered in undertaking this study, however, which emphasizes how difficult it is for older people to stop using psychoactive medications. First, it proved very difficult to recruit subjects into the trial, with 400 of the 493 (81%) eligible subjects declining participation. Further, of the 48 subjects who agreed to participate and were randomised to the psychoactive withdrawal programmes, only 17 (35%) completed the trial. Eight of the 17 subjects who successfully completed the trial also restarted taking psychoactive medications within one month of completion of the study.

In contrast with the benefits of minimising psychoactive medication to lower risk of falling, the evidence of potential benefits of reducing other medications is not as clear. Reduction in anti-inflammatory medication use to prevent falls has been questioned, as this may lead to an increase in arthritic pain and associated reduction in walking speed and general mobility (Bendall et al. 1989; Gibbs et al. 1996). The risks involved with withdrawing antihypertensive medications would appear to far outweigh the potential benefits of falls risk reduction (Stegman 1983), and as such, this approach is not generally regarded as a practical falls prevention strategy. Nonetheless, ongoing monitoring of blood pressure should be undertaken to maintain optimal anti-hypertensive dosage.

5.5.4 Multifaceted Falls Prevention Strategies

To address the issue of multiple risk factors, several studies have now designed and evaluated intervention programmes that involve assessment of risk and subsequent targeting of fall prevention strategies.

Tinetti and colleagues conducted one of the first successful multifaceted intervention programmes (Tinetti et al. 1994). The interventions included medication adjustment, behavioural change recommendations, education and training, and home exercise programmes. During the 1-year follow-up phase, 35% in the intervention group fell, an incidence rate significantly lower than the 47% reported in the control group. This represents a 31% reduction in the incidence of falling. A more recent study in the United Kingdom by Close et al. (1999) reported an even greater falls reduction effect in people who attended an emergency department following a fall. They found that a medical and occupational therapy assessment and subsequent tailored intervention resulted in a significant 61% decrease in fall rates over a 1-year period; however, a similar study conducted in the Netherlands found that multifactorial home visits and subsequent targeted interventions had no effect on falls rates (van Haastregt et al. 2000). This conflicting result may be due to differences in health care delivery between countries.

Another large randomised trial of a multi-factorial falls prevention programme undertaken by Wagner et al. (1994) also reported some benefits of targeted intervention strategies in a large population of older people sourced from a health maintenance organisation. One group received an in-home assessment by a nurse and follow-up interventions targeting inadequate exercise, past falls, alcohol use, medication use, and hearing and visual impairments. A second group received a visit from a general health promotion nurse, and the third group received usual care. The intervention group had significantly fewer falls than the usual care group over the first year of follow-up; however, differences between the nurse assessment with follow-up intervention group and the general health promotion nurse visit group were not significant. The benefits were not well maintained in the second year of follow-up and no difference in falling rates between the groups was found at this time. This suggests the need for ongoing monitoring of and intervention for fall risk factors.

Other falls prevention programmes have used group education sessions. In a randomised trial involving over 3000 independently living health maintenance organisation members age 65 and over, Hornbrook et al. (1994), found that a group education, exercise, and discussion programme reduced falls by 11% — a smaller effect than that reported by the more targeted interventions described previously.

Multifaceted interventions have also been implemented in institutions. Rubenstein et al. (1990) conducted a randomised clinical trial of a specialised post-fall medical assessment (compared with usual care) among 160 ambulatory older people. The assessment involved identification and recommendation for treatment of various fall risk factors (e.g., weakness, environmental hazards, orthostatic hypotension, drug side effects, and gait problems). The 9% reduction in falls among the intervention group over 2 years was not statistically significant, although the intervention group experienced significantly fewer hospital admissions and shorter hospital stays.

A second study in a nursing home setting (Ray et al. 1997) involved the randomisation of seven pairs of nursing homes containing 482 residents at high risk of falls to intervention and control groups. The 221 intervention subjects received individual assessment. There were subsequent recommendations for nursing home staff. These addressed environmental and personal safety, wheelchair use, psychotropic drug use, transferring, and ambulation. The study found a 19% lower rate of falls in intervention facilities.

The available evidence therefore suggests that multifactorial interventions are effective in reducing falls in older people (Feder et al. 2000; American Geriatrics Society et al. 2001; Gillespie et al. 2003). Although it undoubtedly has many advantages, the main disadvantage of the multifaceted approach from a research perspective, however, is that it is difficult to evaluate the relative effects of different programmes and their components. Thus, both factorial study designs and ongoing evaluation of individual programmes are required to establish which components of the multifaceted package are necessary. This area of falls prevention research is changing rapidly. Several trials of this type have been published in recent years, and many more are currently under way. When these findings become available, a clearer picture of effective intervention components and optimal multifactorial approaches will emerge.

5.6 Conclusion

Falling is common in older people, and the consequence of falls can be severe. A number of large prospective studies have determined key risk factors for falling, many of which are amenable to correction. In terms of single risk factor interventions, exercise has been found to be effective, whereas environmental modifications to promote safety in the home and strategies for reducing hazardous medication use have so far been found to be of limited value for preventing falls. Studies addressing further single risk factors amenable to intervention, such as impaired vision, unsafe footwear, and non- or inappropriate use of assistive devices, have yet to be performed. Multifaceted interventions have been found to be effective in preventing falls, particularly those in which older people are assessed by a health professional trained to identify intrinsic and environmental risk factors. Thus, it appears that an understanding of an older person's capabilities and the interaction between the person and the environment provides good scope for falls prevention.

References

Allander, E., Gullberg, B., Johnell, O., Kanis, J.A., Ranstam, J., and Elffors, L. (1996). Falls and hip fracture. A reasonable basis for possibilities for prevention? Some preliminary data from MEDOS (Mediterranean Osteoporosis Study). *Scandinavian Journal of Rheumatology*, (Suppl 103): 49–52.

American Geriatrics Society, British Geriatrics Society, and American Academy of Orthopaedic Surgeons Panel on Falls Prevention (2001). Guideline for the prevention of falls in older persons. *Journal of the American Geriatrics Society*, 49: 664–672.

Avorn, J., Soumerai, S.B., and Everitt, D.E. (1992). A randomized controlled trial of a program to reduce the use of psychoactive drugs in nursing homes. *New England Journal of Medicine* 327: 168–173.

Begg, R. and Sparrow, W. (2000). Gait characteristics of young and older individuals negotiating a raised surface: implications for the prevention of falls. *Journal of Gerontology. Series A, Biological Sciences and Medical Sciences.* 55A(3): M147–M154.

Bendall, M.J., Bassey, E.J., and Pearson, M.B. (1989). Factors affecting walking speed of elderly people. *Age and Ageing* 18(5): 327–332.

Berg, W.P., Alessio, H.M., Mills, E.M., and Tong, C. (1997). Circumstances and consequences of falls in independent community-dwelling older adults. *Age and Ageing* 26(4): 261–268.

Bohannon, R.W. (1997). Comfortable and maximum walking speed of adults aged 20–79 years: reference values and determinants. *Age and Ageing* 26(1): 15–19.

Bond, W.S. and Schwartz, M. (1984). Withdrawal reactions after long term treatment with flurazepam. *Clinical Pharmacy* 3(3): 16–18.

Bray, G. (1995). *Falls risk factors for persons aged 65 years and over in New South Wales.* Sydney, Australian Bureau of Statistics.

Buchner, D.M., Beresford, S.A., Larson, E.B., LaCroix, A.Z., and Wagner, E.H. (1992). Effects of physical activity on health status in older adults. II. Intervention studies. *Annual Review of Public Health* 13: 469–488.

Buchner, D.M., Cress, M.E., de Lateur, B.J., Esselman, P.C., Margherita, A.J., Price, R., and Wagner, E.H. (1997). The effect of strength and endurance training on gait, balance, fall risk, and health services use in community-living older adults. *Journals of Gerontology. Series A, Biological Sciences and Medical Sciences* 52(4): M218–M224.

Cali, C.M. and Kiel, D.P. (1995). An epidemiologic study of fall-related fractures among institutionalized older people. *Journal of the American Geriatrics Society* 43(12): 1336–1340.

Campbell, A.J. (1991). Drug treatment as a cause of falls in old age. A review of the offending agents. *Drugs and Aging* 1(4): 289–302.

Campbell, A.J., Borrie, M.J., and Spears, G.F. (1989). Risk factors for falls in a community-based prospective study of people 70 years and older. *Journal of Gerontology* 44(4): M112–M117.

Campbell, A.J., Borrie, M.J., Spears, G.F., Jackson, S.L., Brown, J.S., and Fitzgerald, J.L. (1990). Circumstances and consequences of falls experienced by a community population 70 years and over during a prospective study. *Age and Ageing* 19(2): 136–141.

Campbell, A.J., Robertson, M.C., Gardner, M.M., Norton, R.N., and Buchner, D.M. (1999a). Falls prevention over 2 years: a randomized controlled trial in women 80 years and older. *Age and Ageing* 28(6): 513–518.

Campbell, A.J., Robertson, M.C., Gardner, M.M., Norton, R.N., and Buchner, D.M. (1999b). Psychotropic medication withdrawal and a home based exercise programme to prevent falls: results of a randomised controlled trial. *Journal of the American Geriatrics Society* 47: 850–853.

Campbell, A.J., Robertson, M.C., Gardner, M.M., Norton, R.N., Tilyard, M.W., and Buchner, D.M. (1997). Randomised controlled trial of a general practice programme of home based exercise to prevent falls in elderly women. *BMJ* 315(7115): 1065–1069.

Cao, C., Ashton-Miller, J.A., Schultz, A.B., and Alexander, N.B. (1997). Abilities to turn suddenly while walking: effects of age, gender, and available response time. *Journals of Gerontology. Series A, Biological Sciences and Medical Sciences* 52(2): M88–93.

Carter, S.E., Campbell, E.M., Sanson-Fisher, R.W., Redman, S., and Gillespie, W.J. (1997). Environmental hazards in the homes of older people. *Age and Ageing* 26(3): 195–202.

Chen, H.C., Ashton-Miller, J.A., Alexander, N.B., and Schultz, A.B. (1991). Stepping over obstacles: gait patterns of healthy young and old adults. *Journal of Gerontology* 46(6): M196–203.

Chen, H.C., Schultz, A.B., Ashton-Miller, J.A., Giordani, B., Alexander, N.B., and Guire, K.E. (1996). Stepping over obstacles: dividing attention impairs performance of old more than young adults. *Journals of Gerontology. Series A, Biological Sciences and Medical Sciences* 51(3): M116–M122.

Clemson, L., Cumming, R.G., and Roland, M. (1996). Case-control study of hazards in the home and risk of falls and hip fractures. *Age and Ageing* 25(2): 97–101.

Close, J., Ellis, M., Hooper, R., Glucksman, E., Jackson, S., and Swift, C. (1999). Prevention of falls in the elderly trial (PROFET): a randomised controlled trial. *Lancet* 353(9147): 93–97.

Connell, B.R. and Wolf, S.L. (1997). Environmental and behavioral circumstances associated with falls at home among healthy elderly individuals. Atlanta FICSIT Group. *Archives of Physical Medicine and Rehabilitation* 78(2): 179–186.

Crowinshield, R.D., Brand, R.A., and Johnston, R.C. (1978). The effects of walking velocity and age on hip kinematics and kinetics. *Clinical Orthopedics and Related Research* 132: 140–144.

Cumming, R.G., Thomas, M., Szonyi, G., Salkeld, G., O'Neill, E., Westbury, C., and Frampton, G. (1999). Home visits by an occupational therapist for assessment and modification of environmental hazards: a randomized trial of falls prevention. *Journal of the American Geriatrics Society* 47(12): 1397–1402.

Donald, I.P. and Bulpitt, C.J. (1999). The prognosis of falls in elderly people living at home. *Age and Ageing* 28(2): 121–125.

Duncan, P.W., Studenski, S., Chandler, J., and Prescott, B. (1992). Functional reach: predictive validity in a sample of elderly male veterans. *Journal of Gerontology* 47(3): M93–M98.

Feder, G., Cryer, C., Donovan, S., and Carter, Y. (2000). Guidelines for the prevention of falls in people over 65. The Guidelines' Development Group. *BMJ* 321(7267): 1007–1011.

Felson, D.T., Anderson, J.J., Hannan, M.T., Milton, R.C., Wilson, P.W., and Kiel, D.P. (1989). Impaired vision and hip fracture. The Framingham Study. *Journal of the American Geriatrics Society* 37(6): 495–500.

Fitzpatrick, R., Rogers, D.K., and McCloskey, D.I. (1994). Stable human standing with lower-limb muscle afferents providing the only sensory input. *Journal of Physiology* 480(Pt, 2): 395–403.

Fleming, B.E. and Pendergast, D.R. (1993). Physical condition, activity pattern, and environment as factors in falls by adult care facility residents. *Archives of Physical Medicine and Rehabilitation* 74(6): 627–630.

Fukuda, T. (1959). The stepping test. Two phases of the labyrinthine reflex. *Acta Oto Laryngologica* 50: 95–108.

Gardner, M., Robertson, M., and Campbell, A. (2000). Exercise in preventing falls and fall related injuries in older people: a review of randomised control trials. *British Journal of Sports Medicine* 34(1): 7–17.

Gerson, L.W., Jarjoura, D., and McCord, G. (1989). Risk of imbalance in elderly people with impaired hearing or vision. *Age and Ageing* 18(1): 31–34.

Ghosh, P. (1985). Aging and auditory vestibular response. *Ear, Nose, and Throat Journal* 64(5): 264–266.

Gibbs, J., Hughes, S., Dunlop, D., Singer, R., and Chang, R.W. (1996). Predictors of change in walking velocity in older adults. *Journal of the American Geriatrics Society* 44(2): 126–132.

Gibson, M.J., Andres, R.O., Isaacs, B., Radebaugh, T., and Worm-Petersen, J. (1987). The prevention of falls in later life. A report of the Kellogg International Work Group on the Prevention of Falls by the Elderly. *Danish Medical Bulletin* 34(Suppl 4): 1–24.

Gill, T.M., Williams, C.S., and Tinetti, M.E. (2000). Environmental hazards and the risk of nonsyncopal falls in the homes of community living older persons. *Medical Care* 38(12): 1174–1183.

Gillespie, L.D., Gillespie, W.J., Robertson, M.C., Lamb, S.E., Cumming, R.G., and Rowe, B.H. (2003). Interventions for preventing falls in elderly people (Cochrane Review). In: *The Cochrane Library*, Issue 4 (Chichester, Wiley).

Gilsing, M.G., Van den Bosch, C.G., Lee, S.G., Ashton-Miller, J.A., Alexander, N.B., Schultz, A.B., and Ericson, W.A. (1995). Association of age with the threshold for detecting ankle inversion and eversion in upright stance. *Age and Ageing* 24(1): 58–66.

Grabiner, M.D. and Jahnigen, D.W. (1992). Modeling recovery from stumbles: preliminary data on variable selection and classification efficacy. *Journal of the American Geriatrics Society* 40(9): 910–913.

Grimley-Evans, J. (1990). Fallers, non-fallers and Poisson. *Age and Ageing* 19: 268–269.

Hasselkus, B.R. and Shambes, G.M. (1975). Aging and postural sway in women. *Journal of Gerontology* 30(6): 661–667.

Hausdorff, J.M., Edelberg, H.K., Mitchell, S.L., Goldberger, A.L., and Wei, J.Y. (1997). Increased gait unsteadiness in community-dwelling elderly fallers. *Archives of Physical Medicine and Rehabilitation* 78(3): 278–283.

Herndon, J.G., Helmick, C.G., Sattin, R.W., Stevens, J.A., DeVito, C., and Wingo, P.A. (1997). Chronic medical conditions and risk of fall injury events at home in older adults. *Journal of the American Geriatrics Society* 45(6): 739–743.

Hornbrook, M.C., Stevens, V.J., Wingfield, D.J., Hollis, J.F., Greenlick, M.R., and Ory, M.G. (1994). Preventing falls among community-dwelling older persons: results from a randomized trial. *Gerontologist* 34(1): 16–23.

Ivers, R.Q., Cumming, R.G., Mitchell, P., and Attebo, K. (1998). Visual impairment and falls in older adults: the Blue Mountains Eye Study. *Journal of the American Geriatrics Society* 46(1): 58–64.

Karlsen, E.A., Hassanein, R.M., and Goetzinger, C.P. (1981). The effects of age, sex, hearing loss and water temperature on caloric nystagmus. *Laryngoscope* 91(4): 620–627.

Kenshalo, D.R., Sr. (1986). Somesthetic sensitivity in young and elderly humans. *Journal of Gerontology* 41(6): 732–742.

Kerrigan, D.C., Todd, M.K., Della Croce, U., Lipsitz, L.A., and Collins, J.J. (1998). Biomechanical gait alterations independent of speed in the healthy elderly: evidence for specific limiting impairments. *Archives of Physical Medicine and Rehabilitation* 79(3): 317–322.

Lawton, M. (1980). *Environment and Aging*. Monterey, California, Brooks/Cole.

Lichtenstein, M.J., Shields, S.L., Shiavi, R.G., and Burger, M.C. (1988). Clinical determinants of biomechanics platform measures of balance in aged women. *Journal of the American Geriatrics Society* 36(11): 996–1002.

Liddle, J., March, L., Carfrae, B., Finnegan, T., Druce, J., Schwarz, J., and Brooks, P. (1996). Can occupational therapy intervention play a part in maintaining independence and quality of life in older people? A randomised controlled trial. *Australian and New Zealand Journal of Public Health* 20(6): 574–578.

Liu, B.A., Topper, A.K., Reeves, R.A., Gryfe, C., and Maki, B.E. (1995). Falls among older people: relationship to medication use and orthostatic hypotension. *Journal of the American Geriatrics Society* 43(10): 1141–1145.

Lord, S., Sherrington, C., and Menz, H.B. (2001). *Falls in Older People: Risk Factors and Strategies for Prevention*. Cambridge, Cambridge University Press.

Lord, S.R. (1990). Falls in the elderly: admissions, bed use, outcome and projections [letter]. *Medical Journal of Australia* 153(2): 117–118.

Lord, S.R., Clark, R.D., and Webster, I.W. (1991a). Physiological factors associated with falls in an elderly population. *Journal of the American Geriatrics Society* 39(12): 1194–1200.

Lord, S.R., Clark, R.D., and Webster, I.W. (1991b). Visual acuity and contrast sensitivity in relation to falls in an elderly population. *Age and Ageing* 20(3): 175–181.

Lord, S.R. and Dayhew, J. (2001). Visual risk factors for falls in older people. *Journal of the American Geriatrics Society* 49(5): 508–515.

Lord, S.R. and Fitzpatrick, R.C. (2001). Choice stepping reaction time: a composite measure of falls risk in older people. *Journals of Gerontology. Series A, Biological Sciences and Medical Sciences* 56(10): M627–632.

Lord, S.R., Lloyd, D.G., and Li, S.K. (1996). Sensori-motor function, gait patterns and falls in community-dwelling women. *Age and Ageing* 25(4): 292–299.

Lord, S.R., McLean, D., and Stathers, G. (1992). Physiological factors associated with injurious falls in older people living in the community. *Gerontology* 38(6): 338–346.

Lord, S.R., Rogers, M.W., Howland, A., and Fitzpatrick, R. (1999). Lateral stability, sensorimotor function and falls in older people. *Journal of the American Geriatrics Society* 47(9): 1077–1081.

Lord, S.R., Sambrook, P.N., Gilbert, C., Kelly, P.J., Nguyen, T., Webster, I.W., and Eisman, J.A. (1994a). Postural stability, falls and fractures in the elderly: results from the Dubbo Osteoporosis Epidemiology Study. *Medical Journal of Australia* 160(11): 684–685, 688–691.

Lord, S.R. and Ward, J.A. (1994). Age-associated differences in sensori-motor function and balance in community dwelling women. *Age and Ageing* 23(6): 452–460.

Lord, S.R., Ward, J.A., Williams, P., and Anstey, K.J. (1993). An epidemiological study of falls in older community-dwelling women: the Randwick falls and fractures study. *Australian Journal of Public Health* 17(3): 240–245.

Lord, S.R., Ward, J.A., Williams, P., and Anstey, K.J. (1994b). Physiological factors associated with falls in older community-dwelling women. *Journal of the American Geriatrics Society* 42(10): 1110–1117.

Lord, S.R., Ward, J.A., Williams, P., and Strudwick, M. (1995). The effect of a 12-month exercise trial on balance, strength, and falls in older women: a randomized controlled trial. *Journal of the American Geriatrics Society* 43(11): 1198–1206.

Luukinen, H., Koski, K., Hiltunen, L., and Kivela, S.L. (1994). Incidence rate of falls in an aged population in northern Finland. *Journal of Clinical Epidemiology* 47(8): 843–850.

Luukinen, H., Koski, K., Laippala, P., and Kivela, S.L. (1995). Risk factors for recurrent falls in the elderly in long-term institutional care. *Public Health* 109(1): 57–65.

MacRae, P.G., Feltner, M.E., and Reinsch, S. (1994). A 1-year exercise program for older women: effects on falls, injuries, and physical performance. *Journal of Aging and Physical Activity* 2: 127–142.

Maki, B.E., Holliday, P.J., and Fernie, G.R. (1990). Aging and postural control. A comparison of spontaneous- and induced-sway balance tests. *Journal of the American Geriatrics Society* 38(1): 1–9.

Maki, B.E., Holliday, P.J., and Topper, A.K. (1994). A prospective study of postural balance and risk of falling in an ambulatory and independent elderly population. *Journal of Gerontology* 49(2): M72–84.

McIlroy, W.E. and Maki, B.E., (1996). Age-related changes in compensatory stepping in response to unpredictable perturbations. *Journals of Gerontology. Series A, Biological Sciences and Medical Sciences* 51(6): M289–296.

McLean, D. and Lord, S. (1996). Falling in older people at home: transfer limitations and environmental risk factors. *Australian Occupational Therapy Journal* 43(1): 13–18.

Morfitt, J.M. (1983). Falls in old people at home: intrinsic versus environmental factors in causation. *Public Health* 97(2): 115–120.

Mulrow, C.D., Gerety, M.B., Kanten, D., Cornell, J.E., DeNino, L.A., Chiodo, L., Aguilar, C., O'Neil, M.B., Rosenberg, J., and Solis, R.M. (1994). A randomized trial of physical rehabilitation for very frail nursing home residents. *JAMA* 271(7): 519–524.

Murray, M.P., Gardner, G.M., Mollinger, L.A., and Sepic, S.B. (1980). Strength of isometric and isokinetic contractions: knee muscles of men aged 20 to 86. *Physical Therapy* 60(4): 412–419.

Murray, M.P., Kory, R.C., and Clarkson, B.H. (1969). Walking patterns in healthy old men. *Journal of Gerontology* 24(2): 169–178.

Nevitt, M.C. and Cummings, S.R. (1993). Type of fall and risk of hip and wrist fractures: the study of osteoporotic fractures. The Study of Osteoporotic Fractures Research Group. *Journal of the American Geriatrics Society* 41(11): 1226–1234.

Nevitt, M.C., Cummings, S.R., and Hudes, E.S. (1991). Risk factors for injurious falls: a prospective study. *Journal of Gerontology* 46(5): M164–170.

Nevitt, M.C., Cummings, S.R., Kidd, S., and Black, D. (1989). Risk factors for recurrent nonsyncopal falls. A prospective study. *JAMA* 261(18): 2663–2668.

Norton, R., Campbell, A.J., Lee-Joe, T., Robinson, E., and Butler, M. (1997). Circumstances of falls resulting in hip fractures among older people. *Journal of the American Geriatrics Society* 45(9): 1108–1112.

Oberg, T., Karsznia, A., and Oberg, K. (1993). Basic gait parameters: reference data for normal subjects, 10–79 years of age. *Journal of Rehabilitation Research and Development* 30(2): 210–223.

O'Loughlin, J.L., Robitaille, Y., Boivin, J.F., and Suissa, S. (1993). Incidence of and risk factors for falls and injurious falls among the community-dwelling elderly. *American Journal of Epidemiology* 137(3): 342–354.

Pearson, M.B., Bassey, E.J., and Bendall, M.J. (1985). Muscle strength and anthropometric indices in elderly men and women. *Age and Ageing* 14(1): 49–54.

Petrovsky, J.S., Burse, R.L., and Lind, A.R. (1975). Comparison of physiological responses of men and women to isometric exercise. *Journal of Applied Physiology* 38: 863–868.

Pitts, D.G. (1982). The effects of aging on selected visual functions: dark adaptation, visual acuity, stereopsis, brightness contrast. In *Aging in Human Visual Functions*. R. Sekuler, D.W. Kline, and K. Dismukes, Eds. New York, Liss.

Province, M.A., Hadley, E.C., Hornbrook, M.C., Lipsitz, L.A., Miller, J.P., Mulrow, C.D., Ory, M.G., Sattin, R.W., Tinetti, M.E., and Wolf, S.L. (1995). The effects of exercise on falls in elderly patients. A preplanned meta-analysis of the FICSIT trials. Frailty and Injuries: Cooperative Studies of Intervention Techniques. *JAMA* 273(17): 1341–1347.

Prudham, D. and Evans, J.G. (1981). Factors associated with falls in the elderly: a community study. *Age and Ageing* 10(3): 141–146.

Ray, W.A., Taylor, J.A., Meador, K.G., Thapa, P.B., Brown, A.K., Kajihara, H.K., Davis, C., Gideon, P., and Griffin, M.R. (1997). A randomized trial of a consultation service to reduce falls in nursing homes. *JAMA* 278(7): 557–562.

Reinsch, S., MacRae, P., Lachenbruch, P.A., and Tobis, J.S. (1992). Attempts to prevent falls and injury: a prospective community study. *Gerontologist* 32(4): 450–456.

Richardson, J.K., Ching, C., and Hurvitz, E.A. (1992). The relationship between electromyographically documented peripheral neuropathy and falls. *Journal of the American Geriatrics Society* 40(10): 1008–1012.

Richardson, J.K. and Hurvitz, E.A. (1995). Peripheral neuropathy: a true risk factor for falls. *Journals of Gerontology. Series A, Biological Sciences and Medical Sciences* 50(4): M211–M215.

Robbins, A.S., Rubenstein, L.Z., Josephson, K.R., Schulman, B.L., Osterweil, D., and Fine, G. (1989). Predictors of falls among elderly people. Results of two population-based studies. *Archives of Internal Medicine* 149(7): 1628–1633.

Robertson, M.C., Devlin, N., Gardner, M.M., and Campbell, A.J. (2001). Effectiveness and economic evaluation of a nurse delivered home exercise programme to prevent falls. 1: Randomised controlled trial. *BMJ* 322(7288): 697–701.

Rosenberg, G. and Adams, A. (1958). Effect of age on peripheral vibratory perception. *Journal of the American Geriatrics Society* 6: 471–481.

Rosenhall, U. (1973). Degenerative patterns in the aging human vestibular neuroepithelia. *Acta Oto-Laryngologica* 76(2): 208–220.

Rubenstein, L.Z., Robbins, A.S., Josephson, K.R., Schulman, B.L., and Osterweil, D. (1990). The value of assessing falls in an elderly population. A randomized clinical trial. *Annals of Internal Medicine* 113(4): 308–316.

Sattin, R.W., Rodriguez, J.G., DeVito, C.A., and Wingo, P.A. (1998). Home environmental hazards and the risk of fall injury events among community-dwelling older persons. *Journal of the American Geriatrics Society* 46(6): 669–676.

Sheldon, J.H. (1963). The effect of age on the control of sway. *Gerontology Clinics* 5: 129–138.

Simoneau, G.G., Derr, J.A., Ulbrecht, J.S., Becker, M.B., and Cavanagh, P.R. (1996). Diabetic sensory neuropathy effect on ankle joint movement perception. *Archives of Physical Medicine and Rehabilitation* 77(5): 453–460.

Skinner, H.B., Barrack, R.L., and Cook, S.D. (1984). Age-related decline in proprioception. *Clinical Orthopaedics and Related Research* (184): 208–211.

Smith, R.D. and Widiatmoko, D. (1998). The cost-effectiveness of home assessment and modification to reduce falls in the elderly. *Australian and New Zealand Journal of Public Health* 22(4): 436–440.

Speechley, M. and Tinetti, M. (1991). Falls and injuries in frail and vigorous community elderly persons. *Journal of the American Geriatrics Society* 39(1): 46–52.

Stegman, M.R. (1983). Falls among elderly hypertensives — are they iatrogenic? *Gerontology* 29(6): 399–406.

Stevens, J.C. and Choo, K.K. (1996). Spatial acuity of the body surface over the life span. *Somatosensory and Motor Research* 13(2): 153–166.

Studenski, S., Duncan, P.W., Chandler, J., Samsa, G., Prescott, B., Hogue, C., and Bearon, L.B. (1994). Predicting falls: the role of mobility and nonphysical factors. *Journal of the American Geriatrics Society* 42(3): 297–302.

Tang, P.F. and Woollacott, M.H. (1998). Inefficient postural responses to unexpected slips during walking in older adults. *Journals of Gerontology. Series A, Biological Sciences and Medical Sciences* 53(6): M471–480.

Teno, J., Kiel, D.P., and Mor, V. (1990). Multiple stumbles: a risk factor for falls in community-dwelling elderly. A prospective study. *Journal of the American Geriatrics Society* 38(12): 1321–1325.

Tinetti, M.E., Baker, D.I., McAvay, G., Claus, E.B., Garrett, P., Gottschalk, M., Koch, M.L., Trainor, K., and Horwitz, R.I. (1994). A multifactorial intervention to reduce the risk of falling among elderly people living in the community. *New England Journal of Medicine* 331(13): 821–827.

Tinetti, M.E., Doucette, J., Claus, E., and Marottoli, R. (1995). Risk factors for serious injury during falls by older persons in the community. *Journal of the American Geriatrics Society* 43(11): 1214–1221.

Tinetti, M.E., Speechley, M., and Ginter, S.F. (1988). Risk factors for falls among elderly persons living in the community. *New England Journal of Medicine* 319(26): 1701–1707.

Tinetti, M.E. and Williams, C.S. (1997). Falls, injuries due to falls, and the risk of admission to a nursing home. *New England Journal of Medicine* 337(18): 1279–1284.

Tinetti, M.E., Williams, C.S., and Gill, T.M. (2000). Dizziness among older adults: a possible geriatric syndrome. *Annals of Internal Medicine* 132(5): 337–344.

van Haastregt, J.C., Diederiks, J.P., van Rossum, E., de Witte, L.P., Voorhoeve, P.M., and Crebolder, H.F. (2000). Effects of a programme of multifactorial home visits on falls and mobility impairments in elderly people at risk: randomised controlled trial. *BMJ* 321(7267): 994–998.

VandenBosch, C., Gilsing, M., Lee, S.-G., Richardson, J., and Ashton-Miller, J. (1995). Peripheral neuropathy effect on ankle inversion and eversion detection thresholds. *Archives of Physical Medicine and Rehabilitation* 76: 850–856.

Vellas, B.J., Wayne, S.J., Romero, L., Baumgartner, R.N., Rubenstein, L.Z., and Garry, P.J. (1997). One-leg balance is an important predictor of injurious falls in older persons. *Journal of the American Geriatrics Society* 45(6): 735–738.

Wagner, E.H., LaCroix, A.Z., Grothaus, L., Leveille, S.G., Hecht, J.A., Artz, K., Odle, K., and Buchner, D.M. (1994). Preventing disability and falls in older adults: a population-based randomized trial. *American Journal of Public Health* 84(11): 1800–1806.

Welford, A.T. (1977). Motor performance. In: Birren, J.E. and Schaie, K.W. (eds). *Handbook of the Psychology of Aging*. New York, Van Nostrand Reinhold.

Wilkins, M. (1999). Health care consequences of falls for seniors. *Health Reports* 10: 47–57.

Winter, D.A. (1995). Human balance and posture control during standing and walking. *Gait and Posture* 3: 193–214.

Wolf, S.L., Barnhart, H.X., Kutner, N.G., McNeely, E., Coogler, C., and Xu, T. (1996). Reducing frailty and falls in older persons: an investigation of Tai Chi and computerized balance training. Atlanta FICSIT Group. Frailty and Injuries: Cooperative Studies of Intervention Techniques. *Journal of the American Geriatrics Society* 44(5): 489–497.

Yip, Y.B. and Cumming, R.G. (1994). The association between medications and falls in Australian nursing-home residents. *Medical Journal of Australia* 160(1): 14–18.

Chapter 6

Epidemiological Approaches to Investigating Causes of Occupational Falls

Tim Bentley and Roger Haslam

CONTENTS

6.1 Introduction

This chapter deals with slip, trip falls (STF), on the level or on steps and stairs, occurring in the workplace. As with other types of accidents, occupational STF are generally complex events. If we wish to understand the causes and identify the most effective means of prevention of this widespread source of work-related injury, detailed investigation and analysis of factors beyond the simple foot–floor interface is necessary. An ergonomics (systems) approach to understanding STF requires the collection and analysis of information relating to a range of individual, equipment, task, environmental, and organisational factors that may contribute to the risk of occupational falls.

This chapter considers approaches to investigating occupational STF at the population level. A major study undertaken by the authors, examining STF among Royal Mail workers in the United Kingdom, is used to illustrate epidemiological approaches to understanding STF causation and prevention (Bentley 1998; Bentley and Haslam 1998; Haslam and Bentley 1999; Bentley and Haslam 2001b). The first section of the chapter describes the use of descriptive epidemiology in the identification of patterns and trends in archival STF records and the *proximal* causes of STF incidents (i.e., those factors that contributed immediately before or at the time of the event). An example of a prospective approach to the collection of data on the immediate causes of accidents, the accident modelling system MAIM, is also presented in this section. The second section describes more detailed investigative approaches, complimentary to descriptive epidemiology. These are beneficial in providing further contextual (e.g., organisational, motivational, cultural) information about STF risk in an organisation and the identification of *distal* causal factors (i.e., latent failures, Reason 1990) contributing to or underlying STF occurrence. Focussed accident-independent survey research and detailed incident follow-up investigation approaches are presented to illustrate how this information may be collected.

6.1.1 Falls in the Royal Mail

At the time the study commenced in 1994, outdoor falls to postal workers represented the largest cause of injury and lost time within the delivery function of the Royal Mail, comprising nearly 30% of reported injuries and over 35% of lost-days. Royal Mail estimated the cost of STF nationally to be approaching £3 million annually in direct costs alone. Moreover, injuries to Postal Delivery Officers (PDOs) are not conducive to employee satisfaction and are disruptive to productivity at the local delivery office level. A major safety priority for the Royal Mail was, therefore, to identify why STF occurred in such numbers and effective countermeasures to reduce their incidence.

A high proportion of postal delivery work in the United Kingdom takes place on foot. PDOs must contend with a range of environmental risk factors during the course of their work, with hazards widespread on customers' premises (Royal Mail delivers mail to the customer's front door) and on public roads and pavements. The relatively uncontrolled nature of the PDO's working environment makes conventional STF prevention measures largely redundant in the mail delivery context. Examples from the Royal Mail study are used extensively in this chapter to illustrate the application of epidemiological approaches to the investigation of falls.

6.2 Epidemiological Approaches: Archival Data Analysis

Because STF accident analysis is concerned with the identification of patterns of work-related STF in terms of who is being injured, where, in what circumstances, and when, descriptive epidemiological methods are used. Descriptive injury epidemiology involves two main activities (Cryer 1995):

1. The description of patterns and trends in occupational injury rates by person (e.g., age, sex, ethnic group), place (e.g., location of accident, region), and time

2. The identification of the causes (or contributory factors) of occupational injury, including both human and environmental factors

Descriptive injury epidemiology has the goals of identifying injury causes, priority areas for preventive action and evaluation of countermeasures designed to reduce incidence. This method of investigation is concerned primarily with the proximal causes of accidents (i.e., those contributory factors immediately preceding the event) (Cryer 1995).

Various sources of archival data may be available to aid STF investigation. In most First World countries, national data is available in the form of official government occupational health and safety statistics, hospital discharge or emergency records, coroner's data (fatalities), research organisation databases, and so on. Most companies also maintain accident-reporting databases. These company databases vary considerably in terms of usefulness of information collected. The analysis of data on PDO STF contained in the Royal Mail accident database is presented to illustrate the uses and limitations of descriptive epidemiological methods and data contained in a typical company database. The aims of the analysis were to:

1. Identify features of the physical and ambient environment commonly present in delivery STF
2. Identify temporal patterns and trends in STF incidence
3. Determine patterns of STF incidence across the PDO population
4. Identify activities at the time of the accident
5. Identify the most common events initiating STF incidents
6. Identify knowledge gaps and further research needs

6.2.1 Method

The database contained information on 1734 delivery STF cases reported between April 1993 and March 1994 in the Midlands division of the Royal Mail. Variables available for analysis (Table 6.1) were typical of company databases and included each accident-involved employee's personal and employment details, details of accident and injuries sustained, time and date, and length of absence from work. A "one-line" narrative description of accident circumstances was also available for each case. Each delivery STF record was manually checked, and erroneous cases (i.e., those that were not STF related) were discarded. Cases with significant amounts of missing information were also withdrawn from the analysis.

Narrative text fields have been acknowledged as important for identifying specific hazards and other aspects of accident circumstances that might otherwise be missed (Stout 1998). As a result of pilot sampling, it

Table 6.1   Information Maintained on Royal Mail's Accident Database

Information Available for Each Case	Examples
1. Accident-involved PDO's delivery office	Leicester North; Woodhouse
2. Cause of accident	"outdoor fall"; "stepped on"
3. Function	Delivery; processing
4. Incident date	June 20, 1993
5. Absence from work (start and end date and total days absent)	June 20–23, 1993; 3 days
6. Nature of injury and body part injured	Sprain/strain; ankle
7. Incident time	7:00 a.m.
8. Age of involved PDO	39 years
9. Sex of involved PDO	Female
10. Length of service of involved PDO	13 years, 3 months
11. "One-line" narrative text field describing accident circumstances	"the employee slipped and fell while walking down an icy driveway"; "the PDO fell down stairs leading to the front door"

was determined that the narrative text provided sufficient information in the majority of cases to allow coding for the following variables:

1. PDO activity at the time of the STF incident
2. Fall initiating event (FIE) (i.e., slip, trip)
3. Underfoot hazard related to fall

A coding form was produced from the pilot exercise, and the narrative text was examined for each case with codes attributed for categories across the preceding variables. Coding was possible for some 86% of cases ("activity"). Once coded, the data were subject to simple frequency distribution, cross-tabulation, and chi-square analysis, using SPSS.

6.2.2 Example Results

A total of 1734 delivery STF cases were distributed across the two years of the analysis: 825 for 1993/1994 (incidence rate: 80.4 per 1000 PDOs) and 909 for 1994/5 (incidence rate: 88.5 per 1000 PDOs). The majority of these incidents did not require a full day's absence from work (61%). More than 3 days absence was required in 25% of cases and over 3 weeks

Table 6.2   Distribution of Delivery STF by Activity, FIE, and Underfoot Hazard

Activity of PDO Immediately before STF	(%)	Fall Initiating Event (FIE)	(%)	Underfoot Hazard	(%)
Walking on the level	63	Foot slipped	49	Ice	46
				Snow	17
				Wet surface or grass	24
				Leaves	6
				Loose surface/ surface movement	6
		Foot tripped	27	Uneven ground	35
				Obstacles	28
				Kerb	28
				Wall/fence	9
		Trod in/stepped on	5		
		Unclassified	19		
Walking on steps (ascending or descending)	19	Foot slipped	61		
		Foot tripped/ missed step	15		
		Other/unclassified	24		
Climbing in/out of vehicle	3				
Climbing on/off bicycle	1				
Unclassified	14				

were required in 9% of cases, confirming that serious injuries are not uncommon for STF occurring during mail delivery work.

6.2.2.1 Activity, Fall Initiating Event, and Surface Condition or Hazard

Table 6.2 gives the distribution of delivery STF by activity of the accident-involved PDO immediately before the STF, the fall initiating event (FIE), and underfoot hazard leading to any slip or trip.

The most common activity was walking on the level between delivery points, although it is uncertain as to how many of these cases may have involved slopes, as steep slopes or inclines were rarely mentioned in the narrative text. Slipping was the most common FIE, accounting for 50% of delivery STF. The proportion of slips is likely to be higher, however, due to the large proportion of unclassified cases. Icy and wet surfaces were

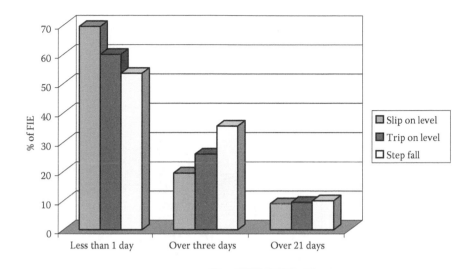

Figure 6.1   Distribution of FIE by days-lost.

the most common underfoot hazards for slipping cases. Tripping was involved in 24% of delivery STF, with three major underfoot hazards identified: uneven ground, obstacles, and kerbs.

Almost one-fifth of the delivery STF occurred while ascending or descending steps, with a slip being the most frequent FIE for step falls (61%). The accident-involved PDO either tripped or missed a step in 15% of cases.

In the absence of severity details for delivery STF injuries, the days absent from work (days-lost) was used as a proxy. The distribution of FIE by days-lost is indicated in Figure 6.1. PDOs more often incurred an injury requiring over 3 days absence from work following a fall on steps (35% over 3 days) compared with PDO who had slipped or tripped on the level. Over 21 days absence was also most common for step falls (10%), suggesting PDOs are more likely to both fall (allowing for exposure) and sustain a serious injury when ascending or descending steps.

The distribution of FIE by body part injured was also considered. Ankle (23%), knee (17%), and back (16%) were the body parts most often injured in delivery STF incidents. Figure 6.2 indicates that the back was the body part most frequently injured for slipping FIE, whereas the ankle and knee were more commonly injured for trip and step falls.

6.2.2.2   Temporal Analysis of Delivery STF

The distribution of delivery STF by month and time of day were considered to determine the impact of seasonal and temporal factors on STF incidence. Figure 6.3 gives the distribution of delivery STF by month.

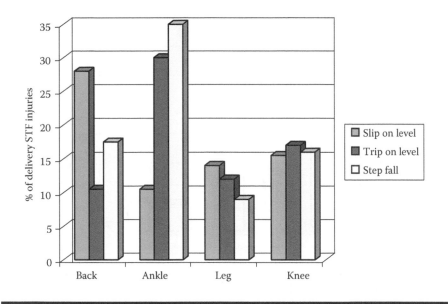

Figure 6.2   Distribution of delivery STF injuries by FIE and body part injured.

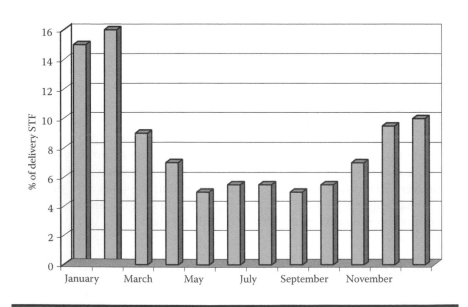

Figure 6.3   Distribution of delivery STF by month.

Table 6.3   Two-Year Incidence Rates (per 1000 Employees) by Sex for All Reported Delivery STF, and 3 and 21 Days-Lost STF Cases

Days-Lost	Male Incidence (per 1000 Employees)	Female Incidence (per 1000 Employees)	Significance
All delivery STF, regardless of number of days-lost	154.4	251.3	P < 0.01
Over 3 days-lost	37.9	72.1	P < 0.01
Over 21 days-lost	11.6	32.5	P < 0.01

The "U" shaped distribution of delivery STF across the calendar year suggests these falls are subject to seasonal variation and winter weather. Indeed, one-half of delivery STF occurred during the four-month winter period November to February.

Further analysis highlighted the impact of winter weather conditions on delivery slip incidents. Analysis of FIE by month indicated that 88% of slips where snow was the underfoot hazard occurred during the month of February (February 1994 + February 1995), with some 80% of snow slips reported during February 1994. A similar cluster effect was identified for ice slips (the most common underfoot hazard in slipping cases). Large proportions of ice and snow slips were found to be concentrated into single days during the winter months.

6.2.2.3 Sex of Accident-Involved PDOs

Incidence rates per 1000 PDOs were calculated and are listed in Table 6.3 by sex and days-lost groupings.

Female PDOs were found to have significantly higher incidence rates than their male colleagues for all reported STF, over 3 days and over 21 lost-days. These data suggest that females are more likely to experience falls and suffer more serious injuries as a result. Further analysis revealed female PDOs had approximately twice the incidence rate for trip falls than males, and were approximately 1.5 times as likely to slip or fall on steps.

6.2.3 Discussion

Slipping was the FIE in one-half of delivery STF, with highest occurrence of slip incidents occurring among older and female PDOs. These findings are consistent with previous studies reported in the falls literature (e.g., Lund 1984; Buck and Coleman 1985), where females and older workers have been found to have the greatest STF risk, although the evidence for these trends in other industries and countries is inconsistent. Slipping was the most common FIE for falls when walking on the level or on steps and was the FIE in nearly all incidents that occurred while climbing in or out of a vehicle. Step falls are also a major concern, with slips on steps (25% of all delivery STF) a high risk for PDOs when the relative exposure to steps as opposed to level ground is taken into account. These data suggest that the interventions likely to be most effective in reducing delivery STF incidence are those designed to reduce the risk of slipping. Furthermore, the large proportion of slips occurring on ice and snow suggests antislip interventions should focus on these risk factors. This is further evidenced by the clustering of falls on ice and snow during a small number of days over the period of the analysis, suggesting minimising exposure to adverse environmental hazards on these days carries the greatest potential for delivery STF prevention.

Tripping was the FIE in approximately one-quarter of cases, with the major underfoot hazards associated with tripping found to be uneven ground (e.g., raised paving slabs), obstacles (e.g., building materials, toys), and kerbs. These findings are in line with previous occupational STF studies (e.g., Andersson and Lagerlof 1983) and research concerned with falls among pedestrians in a public place (e.g., Fothergill et al. 1995). It can be argued that a large proportion of underfoot hazards for both slipping and tripping are avoidable, especially those on householders' premises such as unremoved snow, ice from washing vehicles, toys, building materials, missing handrails for steps, broken footpaths, and other debris and rubbish.

6.2.4 Usefulness and Limitations of Epidemiological Approaches in Fall Investigation Research

The analysis presented previously has demonstrated the potential for epidemiology approaches, using analysis of data on a relatively large number of accident cases, to direct attention to key areas of risk. From the analysis of delivery STF, it was possible to determine incidence rates for different population groups, when STF were occurring across the day, week, and year, as well as the body parts most susceptible to injury. It was also possible to indicate the seriousness and cost of injuries (based on days-lost).

Researchers and safety professionals have often failed to proceed beyond this level of analysis, however, running the risk of failing to identify key risk factors and potential interventions to target these. Through the analysis of narrative text fields in the accident database, the delivery STF study was able to determine the relative contribution of slip, trip, and step falls to STF incidence, together with details of underfoot hazards involved in many of the STF cases. Moreover, it was possible to cross tabulate these with other variables, such as the month and day of accident, to provide detail on the pattern of weather-related occurrence. The finding that ice- and snow-related slip cases clustered around a relatively small number of days per year was an important (if not entirely surprising) finding, and suggests that countermeasures should be focussed on reducing exposure to adverse environmental conditions on the days the hazards are present in a particular area.

A further advantage of the descriptive epidemiological approach is the ability to consider patterns and trends in accidents over time. In the case of the delivery STF study, patterns of STF across the day, week, and year were considered to useful effect. Although the delivery STF analysis only considered data for a 2-year period, other studies in which the authors have been involved have been able to examine trends for STF over several years. This can be useful for fall investigation for a number of reasons:

1. If the impact of annual weather conditions on STF incidence is of interest
2. Where it is necessary to establish the effect of environmental, workplace, behavioural, or organisational change on STF incidence over time
3. In experimental intervention research (e.g., comparing baseline and post-intervention data)

Chapter 9 on STF accidents in the New Zealand logging industry provides a good example of the use of surveillance data in the evaluation of countermeasures for STF prevention at a national level.

Epidemiological approaches also have their limitations. For this reason, it is argued that complementary methods should be used to assist in the interpretation of findings, providing further detail on risk factors and examining the contribution of distal factors, including behavioural and organisational influences.

Limitations of descriptive epidemiological approaches are primarily that:

1. They are dependent on the use of secondary data, with the quality and usefulness of the analysis therefore limited to the quality and quantity of the data that has been collected and collated ("rubbish in-rubbish out").
2. Data may be subject to bias and error at various stages, including problems with reporting (e.g., false reports, inaccurate accounts, poor memory of events); recording (e.g., biased attributions of

blame, misclassification — a "fall" may be recorded as a "struck against" event if the employee fell and hit their head on a wall); analysis and interpretation (bias to fit the investigators' model or philosophy, or inaccuracy, error, and poor judgement in content analysis and coding of narrative data).

3. Analysis typically yields only limited detail on risk factors and the context in which these risks were present.
4. The influence of distal causes and underlying management and organisational factors in STF remain unknown or must be inferred.

Some of these limitations (1. and 3. in the preceding list) are, to a large extent, addressed through the use of prospective approaches to STF epidemiology, where the STF investigator is able to exert control over the type of information collected about fall events and the analysis the data then affords. One such method is discussed briefly next.

6.3 Prospective Epidemiological Methods for Investigating STF

A prospective approach to the collection of data on the proximal causes of STF involves the development and use of modelling systems such as the Merseyside Accident Information Model (MAIM) (Davies et al. 2001a) and the Swedish ISA database (see Andersson and Lagerlof 1983; Kemmlert and Lundholm 1998). The MAIM system has been used to collect information about a wide range of accidents in the U.K. Merseyside region, using a computer-based diagrammatic representation of the sequences of events in incidents, from the first unforeseen event through to an event causing injury (Figure 6.4) (Davies et al. 2001a). Data is collected through accident victim interviews, using an "object-verb-object" notation, through which interviewees are presented with a set of questions and corresponding menus of options that, once selected, are recorded directly in the computer database. The software is programmed so that the interviewee's responses determine future questions. MAIM then structures the accident data so as to facilitate subsequent causal analysis.

Using a prospective approach to data collection, such as with MAIM, has the distinct advantage over archival data analysis of allowing researchers to collect predetermined detailed information on the events and proximal causes associated with STF. Moreover, the use of a computer model provides considerable power and flexibility in hypothesis testing and ease of statistical analysis, allowing complex relationships between variables to be examined. Large data sets can be handled, and both cohort and case-control (e.g., comparison of STF risk factors with those

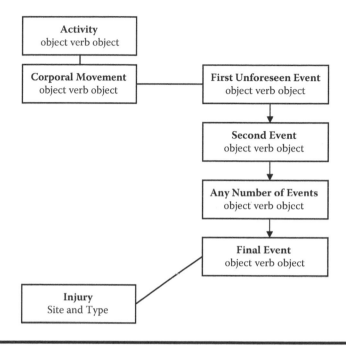

Figure 6.4 MAIM accident model. (Redrawn from Davies, J.C., Stevens, G., and Manning, D.P., 2001a. *Applied Ergonomics*, 32, 141–147.)

for other accident types) study designs can be undertaken. Recent findings from MAIM analyses have confirmed significant associations between carrying items and fall occurrence (Davies et al. 2001a) and falling on steps due to a misstep when wearing bifocal/varifocal spectacles (Davies et al. 2001b). In the later case, the analysis was undertaken following suggestions from researchers (e.g., Howarth et al. 2000) that wearing bifocal/varifocal spectacles on stairs might increase risk of falling (see also Chapter 4).

6.4 Methods and Techniques Complimentary to Descriptive Epidemiology

Beyond employing archival data analysis and prospective accident recording systems, such as MAIM, STF research can benefit from use of other qualitative methods to maximise understanding of the presence and role of STF risk factors in a given situation. This section considers two further forms of primary research, both of which have the general purpose of allowing a better understanding of the role of key risk factors identified through epidemiological approaches. Of particular interest are the influence and impact of distal or latent factors as well as the organisational context in which STF accidents occur. First, accident-independent survey methods are described, including the use of interviews, focus groups, and questionnaire surveys involving a wide range of individuals drawn from the various levels of an organisation — Royal Mail in this instance. Second, the use of detailed accident follow-up investigations with STF-involved employees is demonstrated.

6.4.1 Accident-independent Survey Methods

6.4.1.1 Questionnaire Survey of PDOs and Delivery Office Managers (DOMs)

A questionnaire was administered, comprising a small number of open questions seeking respondents' views on the following issues: unsafe conditions and acts associated with STF, training, attitudes to safety, managers' safety attitudes and activities, and ideas for reducing the incidence of delivery STF. A sample of PDOs (n = 280) and all Midlands DOMs (n = 50) were posted questionnaires. The response rates were 39% (PDOs) and 48% (DOMs), respectively.

6.4.1.2 Interviews with Safety Personnel

Semistructured interviews were undertaken with 17 personnel responsible for safety, including safety managers and advisors as well as local safety representatives. Information collected included details of safety training, training of new recruits, type and quality of footwear and PPE supplied, and management safety activities.

6.4.1.3 Focus Groups

Three PDO focus groups were conducted, with between 8 and 12 PDOs in each group, having a mixed range of experience in delivery work. One local safety representative was present at each meeting. Focus groups were held away from the work area, with participants given a list of issues for consideration and discussion with colleagues prior to the meetings. Information discussed within the three groups included: beliefs on why STF accidents occur; views on how STF might be avoided; safety attitudes and activities of employees and managers; use and quality of footwear and other equipment; and safety training provision and quality.

6.4.1.4 Overview of Results

A large body of information relevant to the problem of delivery STF was generated through use of the accident-independent methods. Examples of the major findings are discussed under the next six headings.

6.4.1.4.1 Adverse Weather Conditions

Adverse weather conditions were described as a major problem for PDOs. In particular, snow and ice make it difficult to achieve performance targets because PDOs have to progress more slowly in slippery conditions. PDOs argued for greater time for deliveries on days where severe weather conditions are present. Slopes and steps leading down to houses were suggested as especially problematic, with householders being understandably poor at clearing away snow and ice from their premises in the early morning.

6.4.1.4.2 Poor Walking Surfaces

Uneven paving, obstacles such as building materials and toys, and holes in walking surfaces were described as commonplace environmental risk factors. These surfaces are most often encountered on homeowners' property. Steps on older properties were indicated as especially hazardous, often lacking handrails.

6.4.1.4.3 Footwear and Equipment

Safety managers and PDOs expressed serious concerns about the type of footwear and frequency of issue to PDOs. Much of the PDO's job is spent walking, and the footwear provided was considered inadequate for such heavy use, with the tread wearing down rapidly. Snow chains, provided to PDOs for working in snowy conditions, were reported as not used by the majority due to their ineffectiveness for most underfoot conditions. PDOs described the chains as being uncomfortable to use on non-snow-covered ground and dangerous on ice. Moreover, drivers could not wear chains because of the time required to put them on and take them off when leaving or entering their vehicle. Trolleys, designed as an alternative to carrying heavy mail pouches, were underused. The main reasons for this was that trolleys slow PDOs down and they could not be used for walks with a high prevalence of steps or flats/apartment blocks. Heavy mail pouches presented an additional loss of balance risk because they were carried asymmetrically. Pouches were reported as particularly heavy on days when magazines have to be delivered. PDOs suggested that heavy mail items, such as magazines, should be delivered on less busy days of the week.

6.4.1.4.4 Unsafe Working Practices

The "job and finish" policy employed by Royal Mail at the time of the study was thought by safety personnel and managers to motivate PDOs to rush and take shortcuts and other risks. PDOs themselves recognised this as a factor and believed that rushing on deliveries was a primary cause of STF accidents. PDOs argued that fast working was necessary to achieve tight delivery time targets, while also agreeing that the job and finish policy provided a strong incentive to hurry and take risks, including carrying overweight mail pouches.

The practice of reading and sorting mail for upcoming deliveries while walking was identified by all participants as an important risk factor. This behaviour reduces the ability of PDOs to identify and avoid underfoot hazards. PDOs noted this practice was widespread because it would take too long to stop and sort mail for the next delivery before walking to the next house.

6.4.1.4.5 Job Training

Training for new recruits was discussed as often restricted to "on the job," involving the new recruit accompanying an experienced worker on his or her delivery route. Both employees and safety personnel believed this practice encouraged the passing on of bad working habits such as taking shortcuts. Safety training was, apparently, not consistently administered to new recruits.

6.4.1.4.6 Management and Organisational Factors

All parties suggested performance, particularly getting the job done quickly, as having priority over safety. The "job and finish" policy and other organisational practices, such as the setting of delivery time targets and the turning of a blind eye to unsafe practices in favour of better productivity, were thought to be powerful influences on employee behaviour. These "behaviour shaping practices" were acknowledged by participants from all levels of the organisation. Other management factors put forward as impacting upon STF risk included the footwear policy and an absence of adverse weather procedures.

6.4.1.5 Usefulness and Limitations of Accident-Independent Methods in Fall Investigation Research

The use of accident-independent survey methods provided a fuller understanding of the risk factors identified from the earlier analysis of STF data. Specifically, additional information was derived from PDOs about the risks associated with walking in slippery conditions, particularly on steep slopes and steps, and where different types of underfoot hazards are most frequently found. For instance, PDOs identified householders' premises, particularly the path leading to the front door where the mail is delivered, as the most hazardous area for obstacles such as toys and building materials. Public pavements in poor repair were suggested as commonplace, with uneven and broken paving a significant hazard for tripping. Information obtained about behavioural factors allowed the researchers to recognise the increased risk these underfoot hazards present when the tasks of walking and examining the mail for the next delivery (i.e., reading addresses and sorting mail) are undertaken in concert. Other unsafe practices identified by the research included rushing and taking shortcuts.

Through the use of these complimentary techniques it was possible to identify the underlying drivers of such unsafe behaviour and the barriers that would have to be overcome for preventative measures to succeed. Organisational policies, such as "job and finish" and delivery targets, were found to serve as "behaviour-shaping factors" that motivate PDOs to take risks implicated in STF causation. Other advantages of accident-independent investigations, as a complimentary method to epidemiological approaches, are those realised through the participatory involvement of workers who would be recipients of change as a result of the findings of the study. The need to understand the problem from the workers' perspective cannot be overstated.

Although the information produced from the accident-independent methods was of considerable value in improving understanding of the STF from a wider, systems perspective, it is desirable to maximise the objective basis for resulting conclusions and recommendations. The final section of this chapter outlines the use of detailed accident follow-up investigations as a form of triangulation with the methods discussed previously.

6.4.2 STF Incident Follow-Up Investigations

If the published literature can be taken as a guide, detailed follow-up investigations as a means of identifying the causes of STF are rarely used. Reasons for this may be many, although the associated cost, practical, and ethical difficulties may, in many instances, make the method unviable. Indeed, the problems involved in setting up such a study can be considerable, with issues to be addressed including: recruitment and access to a representative sample, the practicalities of being informed of accidents and making contact with potential interviewees, and the potential for nonresponse to bias the sample. The expense associated with such a study can be prohibitive, particularly if the sample is large and widely distributed. Many of these problems are less of an issue where the research is conducted in-house in an organisation, with employees of the organisation the subject of investigation.

In whatever context follow-up investigation STF research is undertaken, collection of high-quality data will depend on having well-designed and piloted systematic data collection tools and the willing participation of the study group. The following section provides an account of detailed follow-up investigations undertaken as part of the Royal Mail study.

6.4.2.1 Method

Detailed interviews were undertaken with 40 STF-involved PDOs (Haslam and Bentley 1999). The major purpose of the interviews was to examine the involvement of environmental, behavioural, and organisational factors identified by the accident-independent methods described previously (Section 6.4.1). The investigations took place as soon as possible after the accident was reported (mean interval: 9 days) and at the site of the accident. Respondents were asked to describe in detail the accident events, hazards involved, behaviour, and equipment used. Additional questions dealt with the issues of training and management safety practices. Accident events and contributory factors were recorded on an events and causal factors chart for each case investigated (Haslam and Bentley 1999).

6.4.2.2 Results

To illustrate the method, findings relating to three key areas of concern identified in accident-independent investigations are presented: physical environment, footwear and equipment, and unsafe behaviour and work practices.

6.4.2.2.1 General Details

Ages of involved PDOs ranged from 19 to 56 years, with a mean age of 40 years, 6 months. Experience in job ranged from 1 week to 27 years, with a mean length of service of 9 years, 3 months. Some 22% of the PDOs were female — this percentage being representative of the 18%

female PDO population in the Royal Mail Midlands Division at the time. The injured PDO fell and made contact with the ground in 85% of cases; the remaining PDOs made contact with the door of their van (8%) or stumbled but avoided striking the ground (8%). Injuries were most often to the ankle (35%), most of which were sprains or "twists," and to the knee and lower leg area (23%), most of which were lacerations.

6.4.2.2.2 Physical Environment

Underfoot hazards were most frequently snow (40%) and ice or frost (30%), reflecting the winter period of the study. Damaged paving was the hazard in 8% of cases. "Avoidable" environmental hazards were involved in 23% of cases investigated, with most of these situated on householders' properties.

6.4.2.2.3 Footwear and Equipment

Footwear worn at the time of the STF was inspected and the mean time used for delivery work recorded. The researcher's rating of the tread condition for Royal Mail issued shoes, where the PDO had slipped, was either "very poor" (worn smooth, no tread) or "poor" (little tread remaining) in 75% of cases, although the mean age of the footwear was only 4 months. These findings support those of accident-independent investigations in which PDO and safety personnel argued that footwear issued to PDOs was unsuitable for heavy use and winter conditions.

It was policy at the time of the study for snow chains to be available to PDOs for use in snowy conditions. In none of the cases where slips occurred on snow (n = 16) were snow chains being worn at the time of the STF incident. Reasons given for not wearing snow chains were in line with those raised in the accident-independent investigations.

Where the mail pouch was being carried, it was at least three-quarters full in 40% of cases. In three cases, the weight appeared to have been instrumental in the PDOs overbalancing and falling following a slip. Overweight pouches were being carried in four cases, whereas 77% of PDOs who usually carried their pouches stated it was normal practice to carry overweight pouches. Again, this finding supports those of accident-independent methods.

6.4.2.2.4 Unsafe Behaviour and Work Practices

Table 6.4 provides a breakdown of unsafe behaviours and work practices identified as commonly used by the accident-independent investigations.

Table 6.4   Use of Unsafe Behaviours and Work Practices

Unsafe Behaviour or Practice	PDO Using Practice at Time of Fall (%)	PDO Who Used Practice during Normal Working (%)
Rushing (running, walking very quickly) in slippery conditions	33 (n = 13)	70 (n = 28)
Taking shortcuts (e.g., climbing over a wall/crossing a lawn)	0	65 (n = 26)
Reading or sorting mail while walking	40 (n = 16)	85 (n = 34)
Jumping down steps/skipping steps	0	25 (n = 10)
Jumping off a platform/vehicle	0	10 (n = 4)
Cycle scooting	0	3 (n = 1)
Carrying more than one pouch	2 (n = 1)	20 (n = 8)
Total injured PDOs using unsafe practice at time of STF	60 (n = 24)	

The proportion of STF-involved PDOs who reported using the practice at the time of the incident is presented, together with the proportion of PDOs who stated they usually used the practice during the course of their work. Some 60% of PDOs admitted to using some form of unsafe practice at the time of the STF. The majority of PDOs accepted that using these practices probably increased their risk of having a fall. The reasons given for using these practices included saving time, due to wanting to get home early, time pressures because of delivery targets, and the need to catch a lift or public transport back to the delivery office. These findings supported those of the accident-independent investigations, where sorting mail while walking, rushing, and taking shortcuts were suggested as common practices.

6.4.2.2.5 Risk Factor Interactions

Among this sample of delivery STF incidents, a primary risk factor interaction was the presence of slippery underfoot conditions and the use of unsuitable and worn footwear. The combination of worn footwear, unsafe timesaving behaviour, and slippery conditions were involved in 50% of cases investigated. A second important risk factor combination was the presence of a trip hazard and the practice of reading or sorting mail while walking.

6.4.2.3 Discussion

The brief overview of findings from follow-up investigations illustrates the advantages of primary data collection in addition to the use of archival information in the identification of risk factors for STF. Through the use of detailed investigations, even with a relatively small sample of cases, it was possible to expand on the information produced on proximal factors (i.e., individual, environmental, task-related), and furthermore, provide a more objective basis for the assertions regarding distal factors (e.g., management and organisational influences on PDO work practices), which resulted from accident-independent investigations.

6.5 Conclusions

The triangulation of various accident-centred and accident-independent methods is of considerable value for those seeking to understand the causes of occupational falls at the population level. Although many researchers and safety professionals rely on descriptive epidemiological methods alone, this chapter has argued that useful information, important to understanding the complex factors involved in fall accidents, may be missed where the sole source of evidence is archival accident data. The use of accident-independent investigations, such as surveys of the workforce, can provide information that assists interpretation of the findings of epidemiological work, provides organisational context to epidemiological findings, and identifies the role of distal factors in the accident chain of events. Detailed follow-up interviews provide objectivity, validity, and reliability to the investigator's conclusions regarding the role of proximal and distal factors. These and other aspects to be considered when selecting epidemiological techniques for investigating STF are illustrated in Figure 6.5.

Falls are complex events. Understanding the causes of occupational STF from an ergonomics perspective requires analysis of factors relating to the entire work system, their interaction, and underlying influences. The use of descriptive epidemiological approaches alone may fail to deliver this information, limiting progress toward recommendations for fall prevention.

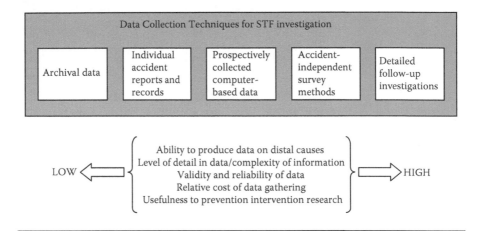

Figure 6.5 Data collection techniques for STF investigation and their relationship to key research concerns.

References

Andersson, R. and Lagerlof, E., 1983, Accident data in the new Swedish information system on occupational injuries, *Ergonomics*, 26, 33–42.

Bentley, T., 1998, Slip, trip and fall accidents occurring during the delivery of mail. Ph.D. thesis. Loughborough University, Leicestershire, United Kingdom.

Bentley, T. and Haslam, R., 1998, Slip, trip and fall accidents occurring during the delivery of mail. *Ergonomics*, 41, 1859–1872.

Bentley, T. and Haslam, R., 2001a, A comparison of safety practices used by managers of high and low accident rate postal delivery offices. *Safety Science*, 37, 19–37.

Bentley, T. and Haslam, R., 2001b, Identification of risk factors and countermeasures for slip, trip and fall accidents during the delivery of mail. *Applied Ergonomics*, 32, 127–134.

Buck, P.C. and Coleman, V.P., 1985, Slipping, tripping and falling accidents at work: a national picture. *Ergonomics*, 28, 949–958.

Cryer, C., 1995, The epidemiology of work-related injury. In: Slappendel, C. (ed.), *Health and safety in New Zealand workplaces* (Palmerston North: Dunmore Press), pp. 15–59.

Davies, J.C., Stevens, G., and Manning, D.P., 2001a. An investigation of underfoot accidents in a MAIM database. *Applied Ergonomics*, 32, 141–147.

Davies, J.C., Kemp, G.J., Stevens, G., Frostick, S.P., and Manning, D.P., 2001b, Bifocal/varifocal spectacles, lighting and missed-step accidents. *Safety Science*, 38, 211–226.

Fothergill, J., O'Driscoll, D., and Hashemi, K., 1995, The role of environmental factors in causing injury through falls in public places. *Ergonomics*, 38, 220–223.

Haslam, R. and Bentley, T., 1999, Follow-up investigations of slip, trip and fall accidents among postal delivery workers. *Safety Science*, 32, 33–47.

Howarth, P.A., Hill, L.D., Haslam, R.A., and Brooke-Wavell, K., 2000, Falls on the stairs — visual risk factors. *Optometry Today*, June 26–27.

Kemmlert, K. and Lundholm, L., 1998, Slips, trips and falls in different work groups with reference to age. *Safety Science*, 28, 59–75.

Lund, J., 1984, Accidental falls at work, in the home and during leisure activities. *Journal of Occupational Accidents*, 6, 181–193.

Reason, J., 1990, *Human error* (Cambridge: Cambridge University Press).

Stout, W., 1998, Analysis of narrative text fields in occupational injury data. In: Feyer, A.M. and Williamson, A. (eds.), *Occupational injury: Risk, prevention and intervention* (London: Taylor & Francis).

Chapter 7

Investigation of Individual Fall Incidents

Paul Lehane and David Stubbs

CONTENTS

7.1 Introduction

Slip, trip falls (STF) in the workplace present particular challenges to those responsible for their prevention and investigation. This chapter examines the issues surrounding this from a psychological point of view, in an attempt to explain the ways in which workplace managers and fall victims might regard STF. It is suggested that certain psychological biases are unhelpful when it comes to falls, in as much as they limit the scope of investigation, reducing and inhibiting effective steps to prevention.

The first author is a practising Health and Safety Inspector for a Local Health and Safety Authority in the United Kingdom, and this chapter is written from the perspective of trying to understand why managers, who are under a legal duty to maintain safety, and indeed some inspectors, seem to have difficulty undertaking meaningful investigations into STF and implementing effective preventative measures.

Research by the authors is outlined in the first part of this chapter, which discusses the ways in which managers view STF and undertake investigations. The second part of the chapter introduces psychological concepts, with subsequent consideration of how these might influence effective investigation.

7.2 STF and Their Investigation

STF are more complex than might be apparent on first consideration. We have developed a framework for fall accidents, which illustrates the wider context (Figure 7.1). The framework identifies five antecedent domains (floor, foot, personal, systems of work, and environmental) and suggests some of the factors that may be associated with each area. All STF involve a loss of balance, and the effectiveness of the recovery reflex determines the outcome and whether injury is likely to occur.

In the United Kingdom, the Health and Safety Executive (HSE) has responsibility for higher-risk work activities, such as construction, farming, and manufacturing, whereas Local Health and Safety Enforcement Authorities have responsibility for lower risk undertakings such as retail operations, leisure activities, warehousing, and office work. Within this sector, Lehane (1998) found that 75% of accidents were investigated by workplace managers as opposed to safety or personnel officers. This places these managers in a key position to influence the prevention of STF. How the managers regard STF will, in turn, influence their behaviour both when undertaking preventative risk assessment and after a fall has occurred. The perceptions of such managers toward STF is an area where little research has been undertaken, but which could provide an opportunity

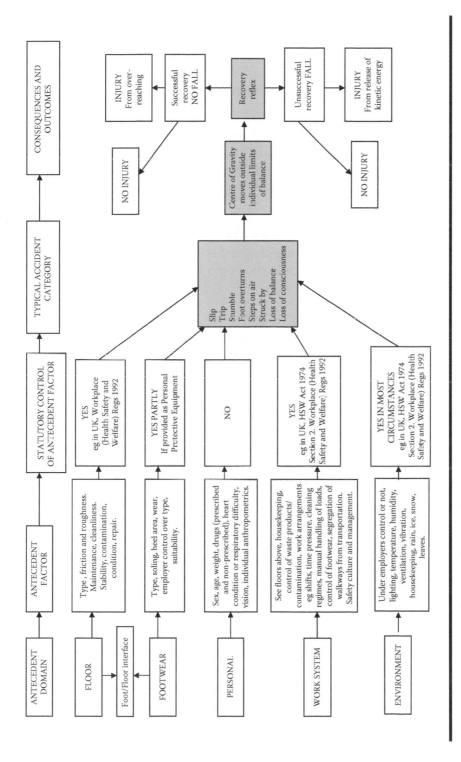

Figure 7.1 Schema to describe slip, trip falls.

for a new approach to fall prevention, given the challenge of reducing numbers of STF below current levels.

7.3 Manager and Fall Victim Perceptions

A study by the present authors into perceptions of STF (Lehane 1998; Lehane and Stubbs 2001) examined 33 incidents. This research focussed on attributions of causal responsibility, foreseeability of the incident by managers, and perceptions as to whether the fall was preventable. It also looked at how managers undertook an investigation and the sources of information utilised.

The 33 STF were all incidents reported under the United Kingdom statutory reporting regulations RIDDOR (Reporting of Injuries, Diseases and Dangerous Occurrences Regulations) (HSE 1995) to Local Health and Safety Enforcement Authorities in the South East of London. The RIDDOR thresholds for reporting are:

- Absence from work of over 3 days
- Major injury (as defined in the Regulations)
- Death

When a suitable accident report was received, the fall victim was contacted for agreement to participate, along with his or her manager. Both were interviewed using a questionnaire developed for the purpose (see Lehane and Stubbs 2001).

7.3.1 Causal Responsibility

Managers and fall victims were asked to rate causal responsibility for the STF into one of three categories:

1. Fall victim/self
2. Other person
3. Both (fall victim/self and other)

The results highlighted interesting differences in the way managers and fall victims perceived slips and trips. Causal responsibility for trips was found to be strongly self-attributed by the fall victim (56% of cases) and to the fall victim by managers (64% of cases). This was not the case for slips, where causal responsibility was self-attributed in only 12% of cases by fall victims and to the fall victim by managers in 37% of cases.

7.3.2 Foreseeability by Managers

From a legal perspective in the United Kingdom, a duty to prevent an accident only arises where the circumstances could reasonably be foreseen. In this sense, the perceptions of managers, regarding the extent to which they could foresee the potential for STF, is important. If managers were to report that STF were easily foreseeable it might be expected to result in a greater propensity to act preventatively.

With regard to the slips in the study, 36% of managers considered that the circumstances could have been foreseen (i.e., 64% did not). The position for trips was even more worrying with only 7% of managers regarding the circumstances foreseeable. The fall victims were asked in turn if they thought their manager should have foreseen the circumstances of their STF. For slips they reported a slightly lower rate than the managers at 29% and a slightly higher rate for trips at 12.5%.

7.3.3 Prevention

In the United Kingdom, a cornerstone of safety management is the requirement under the Management of Health and Safety at Work Regulations 1999 (HSE 1999) for an employer to undertake a risk assessment. The purpose of the risk assessment being to identify potential hazards and ensure that appropriate control measures are in place to reduce risk of injury "as far as is reasonably practicable." An obvious problem exists with STF if, indeed, managers are unable to foresee the circumstances that could lead to a fall.

Given the responses to the question on the foreseeability of STF, it would have been consistent for managers and fall victims to have low expectations for prevention, but this was not the case. For slips, 58% of fall victims and 63% of managers considered them to be preventable, whereas 50% of fall victims and 43% of managers considered trips to be preventable. There was, therefore, an apparent discrepancy among the study participants between being unable to foresee the circumstances that could lead to STF and the belief that STF are preventable. An interpretation of this might be that a tendency exists for managers to make a distinction between the circumstances of a specific accident (foreseeability) and a more general view of accidents elsewhere (prevention). This might operate against managers generalising the findings from an individual accident to prevention on a wider scale.

7.3.4 How Do Managers Conduct Investigations?

Where managers investigate STF, how do they proceed and what factors do they consider? Of the 33 managers interviewed by Lehane and Stubbs

(2001), that had recently had a member of staff involved in a serious STF (17 slips and 16 trips), 23 (69%) of the managers reported having performed some form of investigation. Up to 31%, therefore, were not investigated according to the manager, with 6 of the 10 accidents not investigated being trips. It is perhaps not surprising that one-third of STF were not investigated, given the high level of causal responsibility attributed to fall victims by their managers.

The most common source of information for managers were fall victims themselves, featuring in 22 of the 23 incidents investigated. Other information was collected through observation of the STF location (16 occasions), interviewing witnesses or other people (16 occasions), and from written reports and statements (15 occasions).

Using the antecedent domains in Figure 7.2, investigating managers were asked which, from a list of 22 possible contributory factors, they had considered when carrying out an investigation. Those considered most often were the "type" and "condition of the floor" and the "nature of the footwear." These three factors, relating to the foot–floor interface, accounted for 40% of all factors considered as potentially relevant by managers. Managers considered other factors with the following frequency: personal factors (35%), environmental factors (15%), and the system of work (10%). This is interesting because managers, who represent the employer, are under a legal duty to ensure a *safe system of work*, just as much as they are under a duty to provide a *safe place to work*. On the basis of this, it is suggested that if managers focus on factors relating to the foot–floor interface, proper consideration of other important contributory factors in STF might be prevented.

On this last point, the importance of the interaction between footwear and flooring should not be underplayed; however, it might be useful to draw an analogy with road traffic accidents and the relationship between vehicle tyres and the road surface. It is widely accepted that bald tyres or a worn or contaminated road surface are important factors in the causation of certain types of road accident, but they are not the only relevant aspects. Individual experience of driving would suggest consideration of causal factors ought to be widened to include:

1. Speed of travel
2. Rate of acceleration or braking
3. Loading of the vehicle
4. Weather conditions
5. Lighting
6. The position and actions of other road users
7. Driver skill and experience in the prevailing conditions

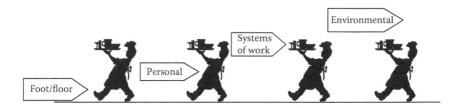

Foot-floor interface	Personal factors	Systems of work	Environmental factors
3 Contributory factors accounting for 40% of factors considered by managers	13 Contributory factors accounting for 35% of factors considered by managers	2 Contributory factors accounting for 10% of factors considered by managers	4 Contributory factors accounting for 15% of factors
• The type of floor • Condition of floor • Type of footwear	• Age • Weight • Height • General health • Eyesight • Smoking habit • Consumption of alcohol • Existing illness or injury • Taking prescribed medicine • Menstruation • Depression • Stress • Level of concentration	• Training • Load being carried	• Day of week • Month • Weather • Lighting

Figure 7.2 Contributory factors for each antecedent domain. (From Lehane, P., 1998, The perceptions and responses of those who experience or investigate occupational slip or trip accidents, M.Sc. dissertation, University of Surrey, Guildford, United Kingdom.)

8. Personal characteristics of the driver (e.g., influence of alcohol or other drugs, tiredness, concentration, visual acuity, reaction time)

From the findings of Lehane (1998) with respect to STF, it is suggested that significantly less attention is paid to personal factors and work systems by managers than to the perhaps more obvious foot–floor interface. These two areas in particular could offer important opportunities for prevention in the same way that improved understanding of personal factors, such as reaction time, concentration, and fatigue, have assisted in the understanding of road traffic accidents.

Lehane and Stubbs' (2001) study was based on STF serious enough to be reported under RIDDOR (HSE 1995). Even so, 31% were not investigated according to the manager interviewed. Given this situation, it is interesting to speculate how many STF occur that do not reach the threshold for reporting and what percentage of these ever receive any investigation. HSC (2003) figures indicate that 10,268 nonfatal major slips, trips, or falls occurring on the same level were reported in the United Kingdom during 2001–2002. Extrapolating the potential number of non-reportable accidents from this using the accident triangles of Heinrich (1931) and Salminen et al. (1992) (referred to in Hale 2001) gives a prevalence of anything between 250,000 and 10 million. The true figure is probably toward the higher end, given that HSC (2003) have estimated that 58% of nonfatal accidents go unreported. A non-investigation rate of around one-third indicates that managers have failed to learn from a very large number of STF.

7.4 Social Psychology and Falls

It is suggested here that three factors are associated with STF, which may result in these incidents being viewed differently from other types of accidents:

1. World knowledge
2. Appreciation of causal mechanisms
3. Psychology and investigation

7.4.1 *World Knowledge*

Since the dawn of time, the human species has had to contend with gravitational force and its effects (Manning 1988). As a biped, humans are inherently unstable and have developed an advanced sense of balance to maintain their vertical stance within certain limits. It is only possible for a person to move a few degrees from the vertical before an automatic recovery reflex is initiated. This involves gross movement of the limbs and trunk in an attempt to bring the centre of gravity back within stable limits. In some situations, it is these gross body movements that lead to some of the typical nonimpact injuries associated with STF (e.g., soft tissue damage such as muscle and ligament strains). Where balance is lost, more serious injuries are likely to arise from the impact of the body against a physical object. The kinetic energy imparted by these falls can be of sufficient magnitude to fracture bone or cause extensive soft tissue injury.

Our everyday experiences have taught us that a loss of balance may result in a fall and possible injury. From the time we take our first steps as a child, falls are an ever-present risk throughout our lives. It may be because of this constant exposure, coupled with the reality that most STF usually occur without serious consequences, that a particular social view of falls has developed. The reaction of parents to a young child when it falls is frequently to play down the significance of the situation, with a response almost dismissive of the event, encouraging the child to ignore what has happened and resume normal activity. These early experiences might go some way to explain the embarrassment that adults often experience when they fall in public; however, young children are better adapted physically to cope with falls than adults. They are shorter in stature and lighter in weight, so the kinetic energy involved is less damaging; their bones have not undergone calcification and do not break so readily. As we age, we lose some of our resilience and the likelihood of suffering injury from falling increases.

It is interesting to speculate that there might even be a deep-rooted "evolutionary" dimension to the way in which society appears to place the onus on individuals to "take care" when negotiating their environment. Consider primitive man out hunting when a member of the hunting party is injured by slipping on some mammoth dung, or trips over a fallen branch while running away from a sabre-toothed tiger (see Figure 7.3). It is likely that the effectiveness of the hunting party would be seriously compromised as a result of these types of accidents. What sort of discussions might have taken place round the cave fire as a consequence? It is easy to imagine that, as a group, our hunters would have learned to

Figure 7.3 Slipping, tripping, and falling: not a new problem. (Owen Williams, Owen Arts Ltd.)

recognise certain situations as being slippery (e.g., mammoth dung) or places where they might trip (e.g., in the woods and undergrowth). Given the difficulty of eliminating such hazards, however, it would be left to the individual to recognise situations that might lead to a risk of falling and act accordingly to avoid them.

We hope this example helps to put the problem of slips and trips into a historical perspective. If we move time forward to the 21st century, what has changed? Has safety management by society advanced from that of our primitive forebears? Humankind still has to negotiate changing terrain and changing conditions. On the one hand, the development of urban living has brought us a myriad of different surfaces to walk on and our collective understanding of the physics of slipping has advanced greatly, both of which have allowed us greater control over floor surfaces. On the other, on the basis of the findings of Lehane and Stubbs (2001), the attitude that it is largely the responsibility of the individual to recognise STF hazards and avoid them, appears to persist outside the specialist fall prevention community. When it comes to having to undertake an investigation of an occupational STF, it is reasonable to assume that an investigator will be influenced by such world knowledge.

7.4.2 Appreciation of Causal Mechanisms

As reported in Section 7.4, Lehane and Stubbs (2001) found that investigating managers attributed responsibility for STF to the fall victim in 64% of incidents. With this in mind, it can be seen how either the existence or absence of causal evidence could lead to managers locating responsibility for STF with the fall victim. In situations where casual evidence is revealed, a frequent conclusion is that the fall victim failed to identify and avoid the hazard. On the other hand, where the investigating manager fails to find any causal evidence, for whatever reason, he or she is left with a fall victim who has been injured in some way, but with no obvious explanation. This may also lead to the causal focus being directed to the fall victim in a way that implies, because no other explanation is available, that the cause of the STF must have been something he or she did or did not do.

In comparison with other accident types, STF may have the appearance of being "low-tech" incidents. They do not usually involve plant, machinery, chemicals or sources of energy (other than kinetic), for example. This perception of STF may itself be a factor in managers not approaching the investigation and prevention with the same vigour as they would for other types of accidents.

Of course STF are actually complex occurrences. One aspect that is sometimes subject to sophisticated investigation in individual slip incidents are the frictional properties of the floor surface, although this probably only happens in a minority of cases. One explanation for this might be that experts with the ability to undertake detailed analysis of flooring are not as readily available as are electrical, mechanical, hydraulic or pneumatic engineers, for example.

7.4.3 *Psychology and Investigation*

In a number of areas, an understanding of psychology might help explain the way in which falls are regarded. Although no specific research has been conducted on the cognitive processes associated with people's response to falls, a significant body of research relating to how we learn from our general experiences is available. Psychological processes, such as "classical conditioning," "operant conditioning," and a range of mechanisms grouped under the term "social learning mechanisms," may be relevant. Although, further consideration of these topics is outside the scope of the present chapter, the interested reader is referred to psychology texts such as Scott and Spencer (1998) and Hayes (2000).

Concentrating on incident investigation, the following sections briefly outline some of the cognitive biases that might influence managers examining STF.

7.4.3.1 *Hindsight Bias*

Hindsight bias results in the outcome of an event, considered afterward, as being viewed as more certain than would have been the case before the event. Greater certainty with hindsight not only affects personal judgement but also perceptions of how an individual believes others should have viewed an event. Thus, a manager undertaking an investigation following an STF incident, subject to hindsight bias, would have an inclination toward the inevitability of the event. The bias could extend further, however, with the manager having an exaggerated view of the fall victim's ability to have foreseen the likely outcome, reinforcing the tendency to place causal responsibility with the victim. It is interesting that the concept of events being "reasonable foreseeable," referred to in the discussion on legal duties with respect to prevention (Section 7.5), is considered "postaccident" with all the attendant distortion of hindsight.

According to Fischoff (1999) being aware of hindsight bias is not sufficient to overcome it. He suggested that you have to force yourself to argue against the apparent inevitability by finding ways in which the outcome could have been different.

7.4.3.2 Confirmation Bias

Ross and Anderson (1991) referred to confirmation bias in the following terms: "Our beliefs influence the processes by which we seek out, store and interpret relevant information ... but an inevitable consequence of our willingness to process evidence in the light of our prior beliefs is the tendency to perceive more support for those beliefs than actually exists in the evidence." Where a manager categorises an accident as either a slip or a trip and recruits an associated causal antecedent, it is likely, according to the theory of "confirmation bias," that any subsequent investigation will tend to focus on information that confirms the causal ascription instead of disproves it.

7.4.3.3 Actor/Observer Effect

The actor/observer effect arises where an observer of a behaviour involving another person (the actor) attributes that behaviour to the actor's disposition, instead of the situation with which they are faced. In reverse, observers can be inclined to attribute their own behaviour to their situation, and not their disposition. For example, the observer of someone slipping might be liable to think that the person should take more care to "look where he or she is going," whereas should they slip themselves, then they may be more likely to blame a hazard in the environment. Again, this bias may tend to direct the attribution of causal responsibility by an investigator toward the fall victim.

7.4.3.4 Self-Serving Bias

The effect of self-serving bias is similar to that of the actor/observer effect in that it will tend to focus causal responsibility toward the fall victim but through a slightly different route. Miller and Ross (1975) cited by Plous (1993) reported that people have a tendency to accept responsibility for good but deny responsibility for bad outcomes. Derived from this, it is possible that investigating managers might steer away from identifying causal factors for STF incidents that reflect badly on themselves, making it more likely that responsibility will be placed with the fall victim.

7.4.3.5 Causal Schema

In the findings reported earlier in this chapter (Section 7.4) from Lehane and Stubbs (2001), it is not clear whether attribution of causal responsibility by managers was made after any investigation had been performed, or whether it derived from an already existing view or causal schema of how

STF occur. It is possible that causal schemas are "learned" from our child and adulthood experiences of falling and are subconsciously recruited as an incident is classified as being either a "slip" or "trip." In this way, an investigating manager might make a judgement as to whether an incident arose from either a "slip" or "trip," then assume causal mechanisms based on his or her previous experiences. This might be compared to the process of stereotyping in which an event or person is attributed to a group sharing certain common features. Once labelled as a member of that group, all its features are attributed automatically without taking into account individual features of the situation.

Woodcock (1996) reported on the recruitment of causal schema by Canadian Safety Officers to explain accidents. Woodcock suggested that such causal schema are developed and refined as a result of exposure to repeated accident investigations. As already highlighted, however, workplace managers investigating STF are unlikely to have had the opportunity to develop and refine causal schema for falls, as they usually will not have the repeated exposure to incident investigation to permit refinement and updating.

Returning to the analogy of a road traffic accident, consideration of an incident involving a car in the early hours of the morning could lead to the recruitment of a schema that such accidents often result from "falling asleep at the wheel." Although this is one possible cause, many others may be overlooked in the face of such a preconception. It would be beneficial if professional investigators, such as police officers, would approach an investigation with an open mind. Indeed, police officers have a range of investigative procedures that they routinely employ to assist this process. For example, driver blood alcohol levels are measured, a detailed examination of the vehicle and the accident scene are performed, and witness statements are collected. It is interesting to compare this approach with the absence of valid and reliable protocols to assist managers when investigating STF.

7.5 Conclusions

This chapter has examined how managers view STF and undertake investigation of such incidents. Workplace managers, with their important influence on safety, have a key role to play in achieving reductions in STF-related injury. Examining and learning from STF incidents might be expected to form an important part of the activities in this respect; however, findings from research on the investigatory practices of managers found that one-third of STF incidents simply were not investigated. With regard to causal responsibility, this was frequently attributed to the fall

Figure 7.4   A holistic approach to consideration of antecedent domains.

victim, especially for trips. Foot–floor interface and personal factors received most attention by investigators; work organisation and environmental factors were considered only infrequently.

It is desirable that the approach to STF incidents moves from the present situation illustrated in Figure 7.2, where a fragmented and unequal consideration of the antecedent domains exists, toward a conceptualisation illustrated by Figure 7.4, where the interdependency of different influences is appreciated and subsumed within a more systemic approach to investigation and prevention. It is no doubt unrealistic to expect workplace managers, investigating STF on an infrequent basis, to achieve this on their own. Tools and methods are needed to support those undertaking STF investigation, in a way that encourages examination of the full range of contributory factors for STF, while overcoming psychological barriers that might constrain attention and preempt objectivity.

References

Fischoff, B., 1999, For those condemned to study the past: heuristics and biases in hindsight. In: *Judgement under uncertainty: heuristics and biases*, edited by Kahneman, D., Slovic, P., and Tversky, A. (Cambridge University Press: Cambridge).

Hale, A., 2001, Conditions of occurrence of major and minor accidents, *Journal of the Institution of Occupational Safety and Health*, 5, 7–21.

Hayes, N., 2000, *Foundations of psychology* (Thomson Learning: London), 3rd ed.

Health and Safety Commission (HSC), 2003, *Health and safety statistics highlights 2002/2003* (HSE Books: Sudbury, Suffolk).

Health and Safety Executive (HSE), 1995, *Reporting of injuries, diseases and dangerous occurrences* (HSE Books: Sudbury, Suffolk).

Health and Safety Executive (HSE), 1999, *Management of health and safety at work regulations 1999* (HSE Books: Sudbury, Suffolk).

Heinrich, H.W., 1931, *Industrial accident prevention* (McGraw Hill: New York).

Lehane, P., 1998, The perceptions and responses of those who experience or investigate occupational slip or trip accidents, M.Sc. dissertation, University of Surrey, Guildford, United Kingdom.

Lehane, P. and Stubbs, D.A., 2001, The perceptions of managers and accident subjects in the service industries towards slip and trip accidents, *Applied Ergonomics*, 32, 119–126.

Manning, D.P., 1988, Slipping and the penalties inflicted generally by the law of gravitation, *Journal of Society of Occupational Medicine*, 38, 123–127.

Plous, S., 1993, *The psychology of judgement and decision making* (McGraw Hill: New York).

Ross, L. and Anderson, C., 1991, Shortcomings in the attribution process: on the origins and maintenance of erroneous social assessments. In: *Judgement under uncertainty: heuristics and biases*, edited by Kahneman, D., Slovic, P., and Tversky, T. (Cambridge University Press: Cambridge).

Salminen, S., Saari, J., Saarela, K.L., and Räsänen, T., 1992, Fatal and non-fatal occupational accidents: identical versus different causation, *Safety Science*, 15, 109–118.

Scott, P. and Spencer, C., 1998, *Psychology — a contemporary introduction* (Blackwell: Malden, Massachusetts).

Woodcock, K., 1996, *Causal reasoning in industrial safety specialists*, Ph.D. thesis, Department of Mechanical and Industrial Engineering, University of Toronto, Ontario, Canada.

Chapter 8

The Assessment and Prevention of Pedestrian Slip Accidents

Steve Thorpe, Paul Lemon, and Stephen Taylor

CONTENTS

8.1 Introduction

Chapter 2 of this book gives a detailed overview of human locomotion and the biomechanics and physics involved in slipping. The present chapter considers the assessment and prevention of pedestrian slipping from the perspective of ongoing efforts at a national level to reduce incidence of fall-related injury in the United Kingdom. This illustrates some of the practical issues arising from attempts to address the problem of slips in the workplace and in public areas.

To many people, slip and fall accidents are perceived as trivial, sometimes humorous, and generally without consequence. This view is reinforced throughout our lives from childhood, where comic book characters slip on banana skins, through to the content of popular television shows, such as "America's Funniest Home Videos" and "You've Been Framed." The victims of the featured slip incidents, however, are generally portrayed as being merely embarrassed, as opposed to sustaining serious physical injuries.

The reality is that pedestrian slips are common in all business and leisure sectors and are a major source of accidents leading to absence from work (Courtney et al. 2001). Statistics indicate that around one in three nonfatal major workplace injuries in the United Kingdom are the result of a slip or trip (HSC 2004), the vast majority of these involving slips. The costs to the United Kingdom economy arising from these accidents are high, estimated at around £1 billion a year. Costs to victims injured through slipping include loss of income, pain, and reduced quality of life. Costs to employers include civil damages, administration and insurance costs, loss of key personnel, and loss of production. Consequences for society include loss of output, medical costs, and social costs.

Furthermore, although official reports indicate that approximately two fatal workplace slip accidents are recorded per year throughout the United Kingdom, we suggest that this is likely to be a gross underestimate for two reasons:

1. Slips are often the first event in falls from height, which account for more fatal accidents than any other cause.
2. Injuries from slipping may lead to fatal complications, such as thrombosis or embolisms, particularly in elderly victims, with the cause of death attributed to the complications instead of the original fall.

In the United Kingdom, the government has established targets for achieving a 10% reduction in the incidence of fatal and major workplace injuries over the next decade (DETR 2000). A considerable reduction in the number of workplace slip accidents will be needed to achieve this. The Health and Safety Executive (HSE) is addressing this through the adoption of slips as one of eight priority areas for enforcement and scientific research. The approach described in this chapter is part of this activity.

8.2 The Assessment of Slipperiness

8.2.1 Site-Based Techniques

Work undertaken by the Health and Safety Laboratory (HSL)* for over a decade has focused on the identification of a robust test methodology for the assessment of the slipperiness of pedestrian walkway materials in workplace areas. The HSE/HSL standard "on-site" slipperiness assessment test methodology is based on the use of a "pendulum" coefficient of friction test (BS 7976 2002a,b,c) (Figure 8.1) and a surface microroughness test (Figure 8.2) performed exclusively using the "R_z" parameter (see Chapter 2, Section 2.7.1). The methodology is supported by HSE Guidance (HSE 2004) and is closely linked to that recommended by the United Kingdom Slip Resistance Group (UKSRG 2000).

It is important to note that, although the coefficient of friction (CoF) and the microroughness of floor surface materials are intrinsically linked to their slipperiness, measurement of CoF and microroughness produces information that should not be considered as independent, direct measures of slipperiness. Slip resistance is neither a constant nor an intrinsic property of any flooring material; contamination, use, and maintenance all play a part in the level of slip risk experienced by pedestrians. Although the two favoured test methods described previously are well established within the United Kingdom and Europe, many other test methods are currently available. Many of these tests, such as manually or self-propelled "sled-type" tests, can be used to make reliable and robust measurements in dry conditions; however, many have reportedly produced unreliable, potentially misleading data in wet or contaminated conditions (Rowland 1997).

It has been reported (Lemon and Griffiths 1997) that the surface microroughness of flooring materials is closely related to their coefficient of friction in wet or contaminated conditions. The relationship between

---

* The Health and Safety Executive (HSE) is responsible for the enforcement of workplace health & safety throughout Great Britain. The Health and Safety Laboratory (HSL) is an in-house agency of HSE and plays a supporting role in delivering HSE's mission to ensure that risks to people's health and safety from work activities are properly controlled.

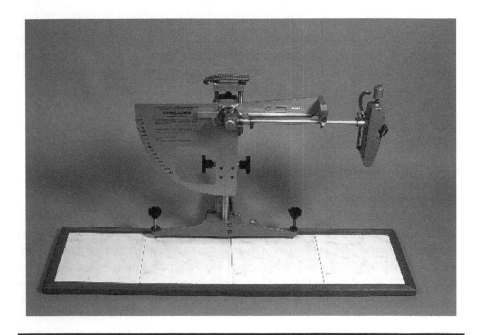

Figure 8.1 **"Pendulum" coefficient of friction test.**

the viscosity of wet contaminants present at the shoe–floor interface and the floor surface roughness required to "break through" contamination has also been studied (HSE 1999; HSE 2004; Lemon and Griffiths 1997). The indication from this is that higher levels of floor surface roughness are required to produce satisfactory friction as the viscosity of floor surface contaminant increases (Table 8.1).

HSE, HSL, and members of the UKSRG favour the use of the "R_z" microroughness parameter for the generation of roughness information relating to floor surface slipperiness. The R_z parameter (formerly known as "$R_{z(DIN)}$" or "R_{tm}", and not to be confused with the "R_z" "ten-point-height" or the "$R_{z(JIS)}$" parameters) represents a mean value of the maximum peak-to-valley height within millimetre-scale sections of floor surface. Use of the R_z parameter has been demonstrated, via a large number of HSL forensic investigations in support of HSE enforcement, to relate closely to pedestrian slip risk in wet conditions.

Measurement of R_z microroughness may be undertaken easily using a "stylus-type" instrument (see Figure 8.2). Recent work by HSL, however, with more complex microroughness analysis techniques, suggests that R_z does have limitations as a measure but, if used in conjunction with other information, gives a valuable indicator of how a floor might perform in wet conditions. Other HSL research (HSE 1999) has demonstrated that measurement of the "R_p" parameter (a measure of the maximum peak

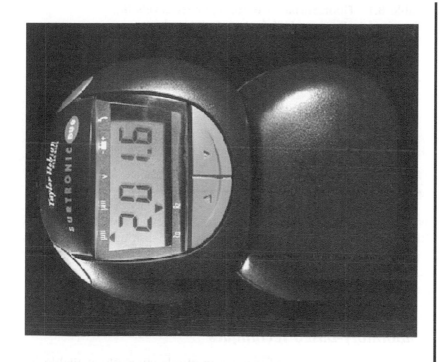

Figure 8.2 Surface microroughness transducers.

Table 8.1   Floor Surface Microroughness Levels (R_z) Required to Produce Satisfactory CoF Levels

Contaminant Viscosity (cPs)	Workplace Analogue	Minimum R_z Floor Microroughness (µm)
<1	Clean water, coffee, soft drinks	20
1–5	Soap solution, milk	45
5–30	Stock	60
30–50	Motor oil, olive oil	70
>50	Gear oil, margarine	>70

height within the analysis length, related to "R_{pm}") allows the production of data that correlate marginally more closely with slipperiness than does the "R_z" parameter. At the time of writing, research is currently underway to produce an optimum "hybrid" surface roughness parameter for the assessment of pedestrian slip risk.

8.2.2 Laboratory-Based Techniques

Despite access to the majority of the test methods currently available for the assessment of floor surface (and shoe sole) slipperiness, the HSE/HSL laboratory-based technique for slipperiness assessment is closely based on the methodology described for "site-based" use. Under most circumstances, pendulum CoF data and R_z microroughness data are used as the basis for assessment. If possible, however, standard practice involves the use of a "subject-based ramp" coefficient of friction test, which may be used to assess the slipperiness of bespoke footwear–contaminant–flooring combinations, as depicted in Figure 8.3. The methodology used by HSL is broadly based on that specified by two German Deutsches Institut für Normung (DIN) standard test methodologies:

1. DIN 51130 (2004), a "shod" test that uses standardised footwear (standardised safety boots manufactured to EN 345), flooring materials, and medium viscosity engine oil as the contaminant.
2. DIN 51097 (1992), a test performed in barefoot conditions using standardised flooring materials, using water (with detergent-based wetting agent) as contaminant.

HSL have developed their own ramp method, using the hardware developed for the two DIN standards outlined previously, to allow the collection of information relating more closely to "real" industrial work-

Figure 8.3   HSL "DIN Ramp" coefficient of friction test.

place situations. The method uses standardised footwear (when used to assess the slipperiness of flooring materials), soled with "Four-S" (standard simulated shoe sole) rubber or "Five-S" (standard simulated soft shoe sole) rubber (also known as TRRL rubber), standardised flooring materials (when used to assess the slipperiness of footwear soling materials), and, under normal circumstances, potable water as contaminant, applied as a fine mist at 6 litres per minute.

8.2.3 Slips Assessment Tool

In addition to the preceding techniques, HSE and HSL have recently developed a PC-based package that allows "nonexperts" to assess the slip risk potential presented by level pedestrian walkway surfaces. This HSE "Slips Assessment Tool" (SAT)* prompts the user to collect surface microroughness data from the test area using a handheld surface microroughness meter. Further information is then fed into the system, such as the floor surface type, the cleaning regime used, the condition of the floor (both in terms of its cleanliness and history), type of footwear worn, and human factors relating to pedestrian use. On completion, a "slip-risk classification" is supplied to the user; this gives an indication as to the potential for a slip. The SAT is designed to assist in the decision-making process when considering the risk of slipping in a defined area; however, it should not be relied upon when considering the performance of a flooring material. In this instance the pendulum method should be used.

In addition, the SAT is also a valuable source of training information, which aims to increase the awareness of the scale of the problem of pedestrian slips, and to familiarise the user with common slip-resistance test methods.

8.3 Contamination of Flooring

It is the interaction of footwear, flooring materials, and contamination that govern slip potential. The vast majority of flooring surfaces present an acceptable slip risk when dry, clean, and free from water, oil, fat, or other slippery substances; however, the slip resistance of pedestrian walkway materials is affected considerably by the presence of fluid contamination. The majority of fluid-based floor contaminants found in the United Kingdom workplace are water-based, although clean water itself is rarely encountered. Water-based contaminants, such as beverages, cleaning solutions, and rainwater, may all pose a greater slip risk than clean water due to increased viscosity.

Contamination with more viscous wet substances (e.g., oil, grease, or detergents), dry materials (e.g., dust, lint, flour, or sand), or semisolids (e.g., fruit, vegetables, or other "wet" products) may greatly increase the risk of slipping. The slip risk posed by floor contamination with substances such as these may be increased to an unacceptably high level (depending on the characteristics of the flooring surface). In such situations, the

---

* The Slips Assessment Tool (SAT) is available and can be downloaded for use free of charge from the HSE website at http://www.hse.gov.uk/slips.

contamination should be removed quickly. If this is not possible, then others working in the area should be alerted to the risk and the area cordoned off until action has been taken to remove the contamination. Similarly, floor coverings that are not suitable for use in wet conditions (due to unacceptably low slip resistance), and that become wet as a result of cleaning, should be cordoned off and allowed to dry completely before pedestrian traffic is allowed to return. Signs used during cleaning should be removed once the floor has dried. Failure to do this will result in their effectiveness as warnings being significantly reduced.

It should be emphasised that the presence of even very small amounts of fluid contamination can significantly reduce the slip resistance of "smooth" floors (i.e., those with an "R_z" surface microroughness of less than around 20 µm). A thin film of water left by an "almost dry" mop, or transferred from a wet area by way of a shoe sole, can lead to a reduction in measured CoF from over 0.70 (dry) to as little as 0.10 (trace wet). Research has also demonstrated that the application of a single drop of fluid contaminant is sufficient to result in a similar increase in slipperiness.

8.4 The Designer's Dilemma

The operational requirements of flooring materials, often considered carefully by architects and floor specifiers, include: appearance, durability, cost, thermal comfort, static loading, chemical resistance, hygiene, electrostatic discharge properties, acoustic properties, impact resistance, abrasion resistance, anti-taint, permeability, fire safety, and level (flatness). The experience of HSL specialists suggests that consideration given by architects and floor specifiers to the slipperiness of flooring materials is often low and, in circumstances where the matter *is* taken seriously, incorrect decisions can be made as a result of potentially misleading data produced by "sled-type" CoF tests (see Case Study 1, Section 8.5.1). Considerable scope exists for the installation of safer flooring in public areas and the workplace.

The measures that can be taken by designers in consultation with their clients, to promote safe pedestrian conditions, need to take into account the full range of contamination that might occur. Furthermore, the cleaning and maintenance regimes necessary to ensure satisfactory "antislip" performance of floor surfaces over the medium to long term ought also to be considered carefully. Designers and specifiers should be aware that the slip resistance of an installed floor might be quite different from that of an "ex-factory" product. The installation and finishing of many different types of flooring, especially ceramic tiles, vinyl, and terrazzo floorings, can cause the surface to change significantly, as can long-term floor surface wear. Examples of installation methodologies that may affect the slip resistance of a floor include the burnishing or polishing of resilient floorings, and the honing of terrazzo floors. It is also necessary that flooring materials receive appropriate treatments after installation and continue to receive maintenance throughout their service life. Floor surface characteristics may change significantly through the lifetime of a floor.

Slipping often occurs where pedestrians encounter unexpected differences in floor conditions. The relative difference between dry CoF and wet (contaminated) CoF characteristics of flooring materials is important. Inside and outdoor environments are routinely dissimilar, and the interface between the two can often lead to problems. This may be the result of ingress of water from external to internal environments, or through temperature changes resulting in condensation on pedestrian surfaces. Measures that may be taken to reduce the likelihood of floor contamination near entrances may include: use of suitable entrance matting systems to prevent the spread of rainwater over internal floorings, use of canopies over entrances, positioning entrances to reduce the effects of prevailing weather, and the use of ventilation systems to help reduce the ingress of wet weather.

Specification and use of the correct type and area of entrance matting is important. Sufficient matting should be used to ensure that shoe soles are dry before the internal flooring is reached. The correct specification of entrance matting systems will not only help reduce incidence of slipping but also reduce cleaning costs and increase the life of a floor. Correct specification (e.g., Figure 8.4) should also prevent the necessity for use of supplementary matting during inclement weather. Suitable matting systems should also be considered inside buildings; for example, within a large shopping mall matting might be used to prevent cross contamination of different areas.

Matting should be cleaned and maintained appropriately and regularly. Poorly maintained matting can act as a sponge, leading to pedestrians stepping from the matting with wet shoe soles. All matting should be fixed securely so that it does not present a tripping hazard. The frequency of cleaning and maintenance regimes, of both floors and entrance matting systems, should be determined by the numbers (i.e., normal and peak levels), and type (i.e., children, elderly, disabled, etc.) of pedestrians who will use the floor.

In situations where it is reasonably foreseeable that installed flooring surfaces are likely to become wet or contaminated, flooring with "enhanced slip resistance" should be specified. The use of such flooring is particularly important on steps and sloping areas and where gradients are used to drain water from areas that will predictably become wet; however, it is not sufficient for specifiers to simply choose a floor covering material with a high dry coefficient of friction value. Such

Figure 8.4 Use of supplementary entrance mats during inclement weather.

floorings often perform badly when wet-contaminated. It is also good practice to avoid using significantly different flooring materials, in terms of their slip resistance, in adjacent areas.

If profiled flooring is specified (i.e., flooring with a pronounced graduated surface), research has demonstrated that the roughness relationships outlined earlier (Section 8.2) still apply for shod pedestrians (Lemon et al. 1999). Depending on the dimensions of the profile, the roughness of the floor may be recorded on both the tops and in the recesses of the profile, although roughness readings taken from the profile tops often relate best to slipperiness in shod, workplace conditions. Such floors will become less slip resistant if contamination is allowed to build up on the surface, which effectively reduces the surface roughness. Profiled floors will generally be more difficult to clean than smooth floors, with difficulty usually increasing with increasing surface roughness or profile height. Research has demonstrated that flooring materials with high surface roughness levels need not compromise hygiene, but may require greater attention to be cleaned successfully (HSE 1996; Richardson 1996). Further guidance on this topic can be found in CCFRA (2002).

Architects and floor specifiers should consider the volumes of pedestrian traffic, particularly peak volumes, when choosing flooring materials.

For instance, speed of pedestrian movement and changes of direction and level will all place extra demands on flooring, such as increased wear. Poor lighting, inside or outdoors, can significantly increase the risk faced by pedestrians, impeding their ability to detect slippery floor conditions. The presence of steps, stairs, or ramps presents an increased risk, and it is important that these features are clearly identified and well lit. Noise in an area can prove a distraction and reduce attention given by pedestrians to the walking surface.

In occupational settings it is often possible to control the footwear that is worn. An informed choice of footwear may offer some protection against slipping. Similarly to floor surfaces, footwear needs to be properly cleaned and maintained to remain effective. In situations where footwear is not controlled, higher priority must be given to other contributing factors. For example, people pulling or pushing loads will require a surface with a higher slip resistance to operate safely; however, workers may not always make allowances for the slipperiness of flooring when deciding whether or not it is safe to push or pull a particular load (Haslam et al. 2002).

Measures that can be taken by owners or occupiers of buildings, to promote safe conditions in service, include identifying potentially wet areas and using an appropriate floor surface for those locations. Effective cleaning maintenance procedures, using appropriate cleaning materials and dressings, need to be established. Confirmation that procedures are actually being implemented at the appropriate frequency should be subject to periodic audit. There may also be benefit in assessing patterns of pedestrian traffic flow and perhaps redesigning tasks to reduce or eliminate the potential for cross contamination. The preceding measures should be complemented by good housekeeping procedures. One way to facilitate this, especially in situations where spills are commonplace, is to leave absorbent materials at conspicuous, accessible points throughout the facility. The resources required to clean up spills, and in training staff to act immediately on discovery of contamination, are far less than those taken up if a serious fall were to occur.

8.5 Case Studies

As stated previously in this chapter (Section 8.2.1), the risk assessment approach used by the Health and Safety Laboratory is based on the measurement of two floor surface characteristics. The use of the test methods is further examined here through discussion of two case studies, outlining real but anonymous slipperiness assessments undertaken by HSL.

8.5.1 Case Study 1: Public Art Gallery

The design and specification of floor surface materials for use in high profile public areas habitually focuses on aesthetic considerations. It is understandable that priority should be given to the visual appearance of such areas, so as to convey the desired impression to visitors. Furthermore, flooring in public areas is often chosen or designed carefully to allow simple and effective cleaning and maintenance. Other factors, such as durability, are taken into account, especially in areas where high levels of pedestrian movement are anticipated. Unfortunately, floor surface slipperiness is seldom considered during the design stage, despite the high incidence of pedestrian slip accidents in both public and workplace areas.

In circumstances where the flooring characteristics discussed previously (i.e., aesthetics, cleanability, and durability) are required, architects/designers often specify (in cost order):

1. Polished natural stone, such as granite or marble
2. Cementitious- or resin-based conglomerate flooring
3. Vitrified or glazed ceramic tiles

The installation of such flooring, which normally involves the use of grinding and polishing techniques during the final stages of installation (for cases 1 and 2), routinely results in the presence of very low levels of floor surface roughness. As discussed previously, low levels of roughness need not necessarily result in increased slipperiness, providing surfaces are maintained in a dry state during pedestrian use; however, contamination of "smooth" (i.e., low roughness) floorings with fluids, such as water or beverages, results in a marked increase in slipperiness.

The involvement of HSL specialist scientists arose following concerns raised about the floor surface installed in a newly built, prestigious art gallery in the United Kingdom. A highly polished, cementitious "terrazzo-type" flooring material had been installed throughout the lower floor of the building, including the kitchens, the café and bar area, the entrance foyer, and all toilet areas.

The flooring had been selected to provide a combination of aesthetic quality, cleanability, and durability. In this case, the architect had acted responsibly by requesting data from the flooring supplier regarding the slipperiness of the floor surface. Data were supplied that suggested that the flooring possessed satisfactory "slip resistance" for use in wet areas; however, the data had been collected using a "sled-type" CoF test. As discussed in Section 8.2.1, research by HSE/HSL and the United Kingdom Slip Resistance Group (UKSRG) has demonstrated that such tests may

Figure 8.5   On-site testing with the pendulum.

yield misleading information when used to assess the CoF of floor surfaces in wet conditions, often significantly overestimating "wet CoF."

On-site testing with the pendulum CoF test (Figure 8.5) and the R_z surface microroughness meter indicated that the flooring, when contaminated with even a small amount of water-based fluid, was unacceptably slippery. Surface roughness values of below 5 μm were recorded (see Table 8.1, which indicates that surface roughness levels of at least 20 μm are required for areas that are not routinely completely dry). In addition, pendulum CoF values of less than 0.10 were recorded; levels of at least 0.35 are required for "low slip risk" classification.

The low surface roughness and CoF values recorded from the flooring were taken as inputs to HSL's standard risk assessment based approach to the assessment of slipperiness. Other factors were considered including:

- The likelihood of floor surface contamination, which is very high in areas such as the café and bar, toilet areas, and foyer
- The large difference between the CoF of the floor surface in wet and dry conditions
- The lack of control of the footwear worn by public visitors
- Use of the flooring by pedestrians of all ages and physical abilities, and the presence of distractions

HSL's interpretation of the results led to the suggestion that a proprietary acid-etching process be used to increase both the CoF and the surface microroughness of the flooring material *in situ*. Subsequently, laboratory-based assessments were undertaken by HSL using pendulum

CoF and surface roughness tests funded by the contractor responsible for the construction of the site, to assess the performance of a range of proprietary etching products. An effective acid-etching treatment was identified, with etching subsequently performed in "high slip risk" areas throughout the Gallery, such as the café and bar and all toilet areas. It was agreed that slip risk in the foyer area could be greatly reduced by the installation of a large absorbent entrance matting system.

Past research by HSE/HSL into the effectiveness of acid-etching products has demonstrated that such products, which are often based on hydrofluoric acid (HF, an acid routinely used to "frost" glass), may produce significant decreases in the slipperiness of hard floor surfaces; however, etching treatments are often expensive, can detract from the aesthetics of flooring materials, and may significantly reduce their service life. Furthermore, little is known about the long-term stability of etched floor surfaces, both in terms of their slipperiness under wet contamination, their stain resistance, and the effects of the HF on grout, expansion joints, and tile adhesive. As a result, it was suggested that HSL specialists return to the site regularly after the etching process had been completed, to assess any changes in the slipperiness of the floor surface over time. Subsequent on-site testing was performed by HSL, which reported a significant reduction in the slipperiness of the floor surface in all high slip risk areas.

8.5.2 Case Study 2: Large-Scale Food Production Facility

Similarities exist between the requirements of flooring in this second case study, concerning a large food production facility and the public art gallery discussed previously. The durability of floorings used in production-type environments is of high importance, as is the case in public areas. The cleanability of any flooring used in the catering industry is a primary consideration. In addition, the customers of food manufacturers are also likely to have expectations with respect to cleanliness, hygiene, and presentation.

In this example, HSL specialists were contracted by a manager of the food production facility to undertake a series of slipperiness assessments on a range of floor surfaces. Before HSL's involvement, a large area of flooring was installed throughout the production, cooking, storage, and transit areas. Because of the requirement for extreme durability, ease of cleaning, and, to an extent, aesthetic quality, a smooth, resin-based material was installed over the existing floor surface. After curing, the surface presented an extremely hard, impermeable layer, which could be cleaned to the standard required using a high-pressure water and detergent jet system. After installation, however, it quickly became apparent that the floor surface posed a significant slip risk to the workforce. Although the risk of slipping could be greatly reduced by elimination of the fluid and product contamination, the working environment made the level of frequent cleaning that would be required to achieve this impracticable.

HSL specialists performed site-based slipperiness assessments using the pendulum CoF test and R_z surface microroughness meter. The results indicated that, in a clean and dry condition, the flooring resulted in a CoF of around 0.65 (low slip potential). On contamination of the floor with clean water, however, CoF results dropped to below 0.20 (high slip potential).

After consultation, it was decided that three small test areas should be treated with different grades of a proprietary antislip coating (i.e., a resin-based matrix containing sharp, millimetre scale, aggregate). Slipperiness assessments were performed by HSL, and it was demonstrated that the three grades performed well in wet conditions, giving CoF results of 0.39, 0.41, and 0.42 (low slip potential), respectively, after cleaning. R_z microroughness tests were not undertaken due to the "macrorough" nature of the flooring.

The pendulum CoF measurements were taken into account in the HSE/HSL risk assessment process, along with other factors including:

- The extreme difficulty in maintaining the flooring in the clean, dry state throughout the working day
- The ability of existing cleaning methods to clean the replacement flooring materials effectively
- Tasks typically performed by operatives using the floor (pushing, pulling, carrying)
- The wear resistance of the antislip coating (this was assessed via a number of repeat site visits by HSL, during which CoF measurements were taken, to allow for changes in CoF to be monitored)
- The potential effectiveness of the three grades of antislip flooring test areas
- The ability of the employer to issue high CoF footwear to all operatives, including the use of high CoF overshoes for temporary staff and visitors

It was agreed that because of the low slip risk presented by the test areas of antislip flooring, walkways, which were clearly marked by high contrast, coloured strips throughout the facility, should have the highest CoF antislip coating applied. Walkway routes were chosen to allow ready access to all parts of the facility (Figure 8.6). Restricting the remedial treatment to walkways significantly reduced the cost of the improvements required to reduce the slip risk to an acceptable level.

Figure 8.6   Antislip pedestrian walkways marked with contrasting borders.

8.6 Conclusions

This chapter has described the HSE/HSL approach to slipperiness assessment, indicating how this operates in practice with two case studies. It has been highlighted that flooring is usually entirely satisfactory when clean and dry; it is where contamination with liquids or other substances arises that slipping becomes a concern. For this reason, it is important that environments, equipment, tasks, and activities are designed and arranged to reduce the likelihood that contamination will occur. Where it is foreseeable that flooring will become wet or otherwise contaminated, the flooring installation should be appropriate to this. Flooring designers and specifiers need to balance the requirements for appearance, durability, cleaning, and safety, making sure they do not neglect the latter.

References

BS 7976-1, 2002a, *Pendulum testers: specification* (British Standards Institution).

BS 7976-2, 2002b, *Pendulum testers: method of operation* (British Standards Institution).

BS 7976-3, 2002c, *Pendulum testers: method of calibration* (British Standards Institution).

Campden and Chorleywood Food Research Association (CCFRA), 2002, *Guidelines for the design and construction of floors for food production areas* (CCFRA: Chipping Campden, Gloucestershire), 2nd ed.

Courtney, T.K., Sorock, G.S., Manning, D.P., Collins, J.W., and Holbein-Jenny, M.A., 2001, Occupational slip, trip, and fall-related injuries – can the contribution of slipperiness be isolated? *Ergonomics*, 44, 1118–1137.

Department of the Environment, Transport and the Regions (DETR), 2000, *Revitalising health and safety* (DETR: London).

DIN 51097, 1992, Testing of floor coverings: determination of anti-slip properties: wet-loaded barefoot areas: walking method ramp test (Deutsches Institut für Normung: Berlin).

DIN 51130, 2004, Testing of floor coverings: determination of anti-slip properties: workrooms and fields of activities with slip danger: walking method ramp test (Deutsches Institut für Normung: Berlin).

Haslam, R.A., Boocock, M., Lemon, P., and Thorpe, S., 2002, Maximum acceptable loads for pushing and pulling on floor surfaces with good and reduced resistance to slipping. *Safety Science*, 40, 625–637.

Health and Safety Commission (HSC), 2004, *Health and safety statistics highlights 2003/2004* (HSE Books: Sudbury, Suffolk).

Health and Safety Executive (HSE), 1996, *Slips and trips: guidance for the food processing industry* (HSE Books: Sudbury, Suffolk), HS(G)156.

Health and Safety Executive (HSE), 1999, *Preventing slips and trips in the food and drink industries— technical update on floor specification* (HSE Books: Sudbury, Suffolk), FIS22.

Health and Safety Executive (HSE), 2004, *The assessment of pedestrian slip risk: the HSE approach* (HSE Books: Sudbury, Suffolk), S&T1.

Lemon, P.W. and Griffiths, R.S., 1997, *Further application of squeeze film theory to pedestrian slipping* (Health & Safety Laboratory: Sheffield), IR/L/PE/97/9.

Lemon, P.W., Thorpe, S.C., and Griffiths, R.S., 1999, *Pedestrian slipping phase 4: macro-rough and profiled floors* (Health & Safety Laboratory: Sheffield), IR/L/PE/99/01.

Richardson, M.T., 1996, *A correlation study between cleanability and roughness characteristics of surfaces* (Health & Safety Laboratory: Sheffield), IR/L/PE/96/7.

Rowland, F.J., 1997, Recent HSE research into the interface between workplace, flooring and footwear. In: *From Experience to Innovation— IEA'97* (edited by Seppälä, P., Luopajärvi, T., Nygård, C.-H., and Mattila, M.) (Finnish Institute of Occupational Health: Helsinki), vol. 3, pp. 402–405.

United Kingdom Slip Resistance Group (UKSRG), 2000, *The measurement of floor surface slip resistance: guidelines recommended by the U.K. Slip Resistance Group* (UKSRG: Radlett, Hertfordshire).

Chapter 9

Occupational Falls Outdoors: Understanding and Preventing Falls in the New Zealand Logging Industry

Tim Bentley, Richard Parker, and Liz Ashby

CONTENTS

9.1 Introduction

Forestry is becoming one of New Zealand's largest industries. New Zealand's planted production forests total approximately 1.8 million hectares and represents 7% of New Zealand's total land mass (New Zealand Forest Owners Association 2004). Forestry operations remain labour-intensive despite the increasing mechanisation of the industry. Approximately 11,000 people were employed in forestry and logging in 2003 (New Zealand Forest Owners Association 2004).

Forestry is among the most hazardous of all industry sectors internationally. In New Zealand, highest work-related morbidity and mortality incidence rates are found in the logging sector (Kawachi et al. 1994; Myers and Fosbroke 1994; Parker et al. 2002). During 2001, some 402 logging injuries were reported in New Zealand, 122 of which required at least one complete day away from work (Parker et al. 2002). This equates to a lost-time injury incidence rate of 16 per million work hours.

The typical New Zealand forestry regime has a number of major stages. Forestry operations, known as silviculture, involve preparation of the planting site, the planting of seedling trees, and pruning and thinning trees at various stages of growth (usually three pruning "lifts" are undertaken at different developmental stages of tree growth). Harvesting, usually referred to as logging, involves the clear felling of trees at around 28 years. Once felled (Figure 9.1) and the branches removed by chain saw (trimming), the trees are dragged or "broken-out" to a clearing known as a skid site (Figure 9.2). Here the trees are assessed for quality and diameter and workers cut the trees into logs — these tasks are known colloquially as "skid work." The logs are then transported by truck to sawmills and ports for export. This chapter is concerned with understanding and preventing slip, trip falls (STF) injuries in logging operations, as research has identified this forestry activity as being of highest risk in this respect.

The Forest Industry Accident Reporting Scheme (ARS) is used by the Centre for Human Factors and Ergonomics (COHFE) (formally the Human Factors Group of the Logging Industry Research Organisation) to inform its New Zealand forest industry injury prevention research and develop-

Figure 9.1   Felling a 28-year-old plantation pine tree with a chain saw.

Figure 9.2   Skid site where trees are assessed and cut into logs of differing quality grades.

ment programme (Bentley et al. 2002). The scheme covers both silvicultural and logging sectors. The ARS, which has been in existence for approaching 20 years, contains details of lost-time, minor (less than one full day absence from work) and near miss incidents. The New Zealand forest industry strongly supports the scheme, resulting in the large majority of reported forest injuries being included in the database. In turn, COHFE provide quarterly and annual summary injury data to forest company contributors and the wider audience in New Zealand, along with research reports and general injury prevention information. Most important from a research perspective, the injury data received from forest companies enable COHFE to examine trends and patterns in logging injury data, and to target their wider ergonomics, safety, and health programme at key risk areas. The scheme also provides baseline and control data and other evaluation measures helpful in intervention research.

The first part of this chapter presents a descriptive epidemiological analysis of STF injury data reported to the logging ARS during the period January 1996–June 2001. The purpose of the analysis was to provide an improved understanding of STF events and their causes, allowing informed consideration of potential interventions to reduce the risk of injury. The chapter concludes with an overview of a field trial of spiked boots, examining their effects on logger safety, productivity, and workload.

9.2 Epidemiological Analysis of Logging Injury Data

The aims of the analysis were to:

1. Examine patterns and trends in STF injury data, including employee groups at most risk, seasonal effects, and temporal patterns
2. Identify common STF injury mechanisms
3. Identify most frequent fall initiating events (FIE)
4. Identify common underfoot hazards associated with STF injuries
5. Explore possible areas for preventive action
6. Identify knowledge gaps and further research needs

9.2.1 Method

ARS data on lost-time (i.e., the injured logger lost at least one full day of work) and minor logging STF injuries (less than one full day's absence from work) were selected for the period January 1996–June 2001 from the main ARS database. As no data field currently exists in the database for event type (i.e., "slips on the level", "trips"), STF incidents were retrieved from the ARS database via a keyword search of event descriptions using the keywords: "fell down", "fell over", "slip", "trip", "fall over," or "fall down." The cases produced from this search were then manually checked and cleaned, and erroneous cases (i.e., those where keywords were present but did not refer to STF incidents such as "the chain saw slipped ...") discarded. This produced a total STF injury dataset of 318 cases for analysis. Data were analysed using SPSS version 10. Variables considered for analysis are listed in Table 9.1.

Table 9.1   Variables Considered in the Analysis of Lost-Time Skid Site Injuries

Variable	Example
Lost time	3 days; 4–7 days; 8 days and over
Time of day injury occurred	1400 hours
Month and year of injury	August 2000
Part of logging operation	Skid work; felling; trimming
Experience in task	15 months
Body part injured	Foot; lower back
Injury type	Fracture; sprain/strain
Treatment	Doctor; hospital
Fall initiating event	Slip on the level; slip from height; trip
Underfoot STF hazard	Log; branch; vehicle

Quantitative analysis methods were simple frequency distributions, cross-tabulations, and chi-square. Content analysis was performed on one-line narrative text describing injury circumstances to obtain information on two variables: "fall initiating event" and "underfoot STF hazard." Inter-coder reliability was determined for a 20% sample of cases for the fall initiating event and underfoot hazard variables, with inter-rated agreement of 96% fall initiating event (FIE), and 98% underfoot hazards, respectively.

9.2.2   Results

A total of 318 STF injury cases were reported to the logging ARS during the period January 1996–June 2001. Of these, 155 (49%) were lost-time injuries and 163 (51%) were minor injuries. Because "near miss" events are reported inconsistently, near miss cases were not included in the analysis. The 318 STF injury cases resulted in a total of 1288 lost workdays (this figure does not account for the large amount of lost time that will have resulted from minor injuries where only a part day was lost), at an average of 8.35 days lost per case. Some 90 (28%) STF injuries involved lost-time of 4 days or more, and 13% of cases involved 8 days or more lost-time, suggesting a significant proportion of logging STF injuries are of a moderately serious to serious nature. Injured loggers attended a hospital for treatment in 61 cases (19%), saw a doctor in 130 cases (41%), and were administered first aid in 25 cases (8%). Treatment was unrecorded or no treatment was required in 101 cases (32%).

Table 9.2 lists the distribution of logging STF injuries across the five-and-a-half years of the analysis. The number of STF injuries reported each

Table 9.2   Distribution of Logging STF Injuries, Compared with All Logging Injury Types, by Year of Occurrence

	1996	1997	1998	1999	2000	2001 (to end June)	Total
STF injuries (minor and lost-time)	51	43	42	78	58	45	317
All logging injuries	324	275	262	388	386		
STF injuries as a proportion of all logging injuries	15.7	15.6	16.0	20.1	15.0		

year are reasonably consistent, with the exception of 1999 in which markedly more falls were reported and falls contributed a significantly greater proportion of all logging injuries (20.1%). The increase in reported falls in 1999 may have been due to the introduction of quality control tasks in the logging process. On average, STF injuries accounted for 16.5% of all logging injuries over the period of the analysis.

9.2.2.1 Logging Operation

The logging operations in which STF injuries occurred most frequently were skid work (n = 71; 22%) and felling (n = 67; 21%) (Table 9.3).

A number of the other operations listed in Table 9.3 also commonly occur in skid sites (e.g., trimming), suggesting the skid site area to be the most hazardous for STF injuries in logging, and the most immediate target for intervention. Injuries in the skid work part of the operation tended to incur greatest harm (when measured by lost-days) and subsequent expense to the industry, with 34% of cases requiring 4 or more days absence from work, and 17% of cases involving 8 or more days lost-time. Trimming was not far behind in terms of lost-time per injury, with 29% of cases requiring four days or more lost-time and 15% greater than 8 days.

Table 9.3   Distribution of STF Injuries Across Logging Operation

	Skid Work	Felling	Trimming	Breaking Out	Loading	Maintenance	Other
n	71	67	44	44	11	17	64
%	22	21	14	14	3	5	20

9.2.2.2 Fall Initiating Events (FIE) and Underfoot Hazards

Slips from a height (mostly slips from a log or log stack), were the largest category of FIE (n = 135; 42%), followed by slips on the level (n = 101; 32%) and trips (n = 34; 11%). A further 28 cases (9%) were slips that could not be classified as slips on the level or from a height from the information provided in the narrative text. Table 9.4 lists the distribution of FIE, together with underfoot hazards as they relate to each FIE.

"Log/stem" was by far the most common underfoot hazard associated with STF from a height (n = 88; 65% of falls from a height), being the underfoot hazard from which the logger slipped in some 34% of all logging STF injury cases (n = 108). Other underfoot hazards mentioned with relatively high frequency included "vehicle/machine" (n = 39), steep terrain (n = 22) and branch/stick (n = 24).

Slipping from a height also resulted in greatest lost-time, with 21 (16%) of falls from a height requiring 8 or more days lost-time, compared with 8 (8%) for slips on the level, and 3 (9%) for trips. These findings suggest that slipping from a height, particularly slipping while standing and walking on logs, as the most promising target for intervention to impact on the logging STF injury problem. Slips on the level were associated with a wider range of underfoot hazards, the most common of which were steep terrain, branches, and "slash" (the woody debris of logging operations).

The analysis also considered the distribution of FIE across the major parts of the logging operation. Table 9.5 highlights three significant STF injury risk areas:

1. Slips from a height (usually logs or log stacks) on skid sites (skid work and trimming)
2. Slips on the level in felling
3. Slips on the level in breaking out

Skid sites are highlighted as the area of greatest risk for falls from a height, with some 61% of STF injuries occurring on skid sites involving a slip from a height. Interventions to reduce the risk for this fall risk should include consideration of antislip footwear when working on logs, and work practices that reduce the exposure to this risk.

Falls during felling and breaking out also require focused research attention. Common underfoot hazards for STF injuries during felling and breaking out were identified as "log" (n = 19; 17% of felling and breaking out injuries); steep terrain (n = 16; 14%); branches (n = 13; 12%); and slash (n = 10; 9%).

Table 9.4 Distribution of Fall Initiating Events by Major Underfoot Hazards

Fall Initiating Event (FIE)	Log	Branch	Slash (branch broken from a tree)	Steep	Slippery Surface (mud, oil, etc.)	Vehicle	Other	Total
Slip on the level	6	9	14	22	5	0	45	101
Slip from a height	88	5	0	0	0	35	7	135
Slip (mechanism unspecified)	7	1	0	0	2	0	18	29
Trip	5	8	3	0	0	0	19	35
Other	2	1	1	0	0	4	10	17
Total	108	24	18	22	7	39	99	317

Table 9.5 Distribution of Fall Initiating Events by Logging Operation

Fall Initiating Event (FIE)	Skid Work	Felling	Trimming	Breaking Out	Loading	Maintenance	Other
Slip on the level	12	38	6	25	1	4	14
Slip from a height	43	13	22	10	8	9	30
Slip (mechanism unspecified)	4	8	9	0	1	1	7
Trip	10	8	5	6	1	2	3
Other	1	0	2	3	0	1	10
Total	70	67	44	44	11	17	64

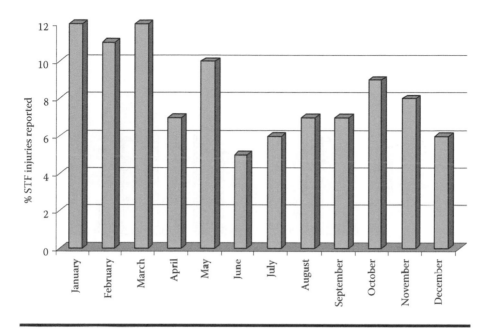

Figure 9.3 Distribution of logging STF injuries reported by month.

9.2.2.3 Seasonal Effects and Temporal Patterns

Figure 9.3 reveals that over one-third of reported STF injuries occurred during the hottest summer months, January to March. The months with lowest STF injury counts were the main winter months of June, July and August. This figure is perhaps surprising when considered alongside findings from other "outdoor" industry studies (e.g., Bentley and Haslam 2001), where falls were generally more common in winter due to slippery underfoot conditions.

A number of factors may be important in explaining these seasonal differences in logging STF injury data. First, the heat in the summer months may be instrumental in logger fatigue and dehydration, factors known to be associated with poor hazard identification and increased injury risk (see Bates et al. 2001; Lilley et al., personal communication). Second, a large proportion of logging work takes place in the North Island of New Zealand where snow and ice are comparatively rare. Indeed, only a small proportion of STF injuries were reported to have involved slippery underfoot conditions, and only one report mentioned snow/ice. Cross-tabulation with FIE indicated no marked effects of season upon slip risk, although falls from a height occurred in notably greater proportion during the month of March (n = 21; 55% of STF injuries during March). Finally, a crude indicator of underfoot conditions is provided by the variable

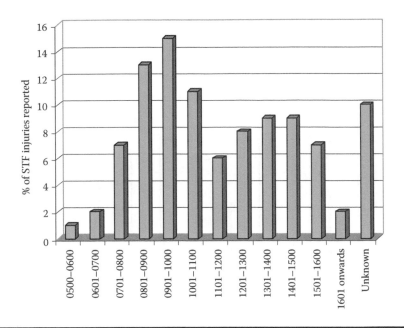

Figure 9.4   Distribution of STF injuries during the workday.

"wet"/"dry." Where this information was provided (n = 188 cases), conditions were noted as "dry" in 36% of cases and "wet" in 24% of cases, suggesting that it is factors other than wet underfoot conditions that are most important in these forestry incidents.

STF injuries were reported with greatest frequency for the first four working days of the week (Monday–Thursday), with the largest proportion on Monday (n = 79; 25%) and Tuesday (n = 61; 18%). These figures are in line with those for all logging injuries (Evanson et al. 2001), and are largely a function of exposure (i.e., increased time spent at work and exposed to STF hazards).

STF injuries were reported in greatest prevalence during the period 0800–1000, with a peak during 0901–1000 (Figure 9.4). These figures are in line with those for all logging lost-time injuries (Parker et al. 2002), and may be due to fatigue (Kirk and Paterson 1996) and dehydration (Bates et al. 2001) before a meal break.

9.2.2.4 Job Experience of Injured Logger

Data for an injured worker's age was not available for this analysis. Examination of experience data, however, produced 91 cases (29%) where the data for experience was available, which involved an injured logger with 0–6 months' experience in his or her job (Table 9.6).

Table 9.6 Distribution of STF Injuries by Injured Logger's Experience in Their Job

Injured Logger's Experience	n	%
0–6 months	91	29
7 months–1 year	30	9
13 months–2 years	27	9
25 months–5 years	62	19
61 months and over	59	18
Unknown	48	15
Total	317	100

A total of 38% had 1 year or less experience. Although no reliable data is available about the proportion of loggers working in the various categories of experience from which to determine exposure, these figures suggest the new recruits and inexperienced loggers are more likely to incur a STF injury than their more experienced colleagues. Further analysis revealed inexperienced (0–6 months) loggers reporting injuries were most commonly engaged in the following logging operations at the time of the accident: felling (n = 27); skid work (n = 15); breaking out (n = 12); and trimming (n = 10).

9.2.2.5 Nature and Seriousness of Injury

Table 9.7 gives the distribution of injury type by severity (as measured by days-lost). Lacerations (n = 96; 30%), sprains/strains (n = 83; 26%) and contusions (n = 66; 21%) were most commonly reported. Injury types resulting in the greatest lost-time were found to be fractures (75% 4 or

Table 9.7 Distribution of STF Injuries by Injury Type and Days-Lost Category

Injury Type	n	%	Less Than One Day Lost (%)	Four or More Days Lost (%)	Eight or More Days Lost (%)
Fracture	28	9	11	75	64
Laceration	96	30	49	26	13
Sprain/strain	83	26	35	37	10
Contusion	66	21	79	5	0
Dislocation	9	3	33	33	22
Other/unknown	36	11			
Total	317	100			

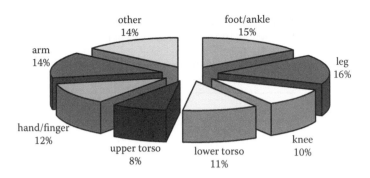

Figure 9.5   Distribution of body parts injured in logging STF incidents.

more days lost-time; 64% 8 or more days-lost) and sprains/strains (37% 4 or more days lost).

The body regions most commonly injured were the upper and lower limbs, with a total of 41% of injuries sited at the lower limb and 31% the upper limb (Figure 9.5).

Analyses conducted in other industries and countries have found a higher proportion of ankle injuries in STF injury data than reported in this study (e.g., Bentley and Haslam 2001). It is possible that the ankle support provided by loggers' boots is contributory to the prevention of injuries to the ankle.

Attempts to break falls using hands and arms may be the reason for the large proportion of injuries in these areas. Training in safer falling methods (i.e., breaking a fall while protecting vital areas and minimising impact on the hands and wrists) may assist in reducing injuries resulting from falls.

9.2.3 *Discussion of Key Findings and Implications for STF Injury Prevention*

The proportion of logging injuries involving STF was fairly consistent across the five-and-a-half years considered in the analysis, with STF injuries contributing significantly to the human and financial cost of logging injuries in New Zealand (16.5% of logging injuries). Indeed, STF injuries are second only to "struck by" incidents, the largest category of logging accident by some margin (Evanson et al. 2001). Some 1288 workdays were lost as a result of STF injuries for the period of the analysis; however, this figure considerably underestimates the true cost of STF injuries because minor injuries will always involve some disruption of work or lost-time during the day of the incident.

Slips from a height occurred with greatest prevalence, with slipping while walking or standing on a log or log stack the major injury mechanism. These events also resulted in largest numbers of serious injuries as measured by days lost. Preventive efforts should revisit the issue of spiked-soled boots (Kirk and Parker 1994) to increase slip resistance, with further consideration given to their effectiveness on different underfoot surfaces (particularly logs), and factors impacting on their use by loggers (e.g., availability and cost). Companies and contractors might also consider the organisation of work and skid-site layout as a means of reducing exposure to the risk of slipping from logs and log stacks and the subsequent fall from a height. Where possible, skid workers should avoid walking or standing on logs. Slips on the level were mostly due to debris around the skid site and felling areas. Improved housekeeping and slash and tree debris removal around landings would assist in reducing the risk of these injuries.

Unlike other studies that have found outdoor STF incidence to increase sharply in winter, this analysis found injuries occurring more frequently during the hottest New Zealand summer months, January through March. It has been suggested that dehydration might be a factor in this, as dehydrated workers have reduced information processing capabilities and subsequent elevated risk of injury. COHFE are currently engaged in a substantive research project examining dehydration among forestry workers, with initial findings indicating that many loggers become clinically dehydrated during the course of a working day. On the basis of this, it is argued that work scheduling should make allowance for breaks, giving opportunity for recovery and fluid/food intake. This assertion is supported by the distribution of STF injuries throughout the working day, with largest numbers occurring during 9–10 a.m., after several hours of work, often without a rest period.

Loggers at most risk are those with least experience in their job and those working in skid work and felling operations. Loggers incurred some 38% of STF injuries in their first year of employment, and 29% had been in their job for 6 months or less. These findings reinforce those from other COHFE research in New Zealand logging and silviculture (e.g., Parker and Bentley 2000; Ashby et al. 2001), and suggest that the greatest return in terms of injury prevention would be expected from targeting countermeasures in this worker group.

Training in STF injury prevention, including improved hazard identification, assessment and control, safe working methods, footwear requirements and protecting vital areas (i.e., the head and neck) in the event of a fall, may be the most effective method for STF injury reduction. This is because countermeasures commonly recommended for STF prevention in other industries (e.g., housekeeping, flooring) are not as relevant to the logging industry, where the working environment is uncontrolled when compared with indoor workplaces. In addition, loggers normally wear heavy and restrictive personal protective clothing (e.g., helmet, earmuffs, visor, high visibility vest or shirt, chain saw cut-resistant leg wear, steel capped boots) and carry a 10-kg chain saw and wear a gear belt containing a hammer, wedges, and a fire extinguisher, which can contribute to the loss of safe footing and difficulty regaining balance once it is lost.

The impact of one countermeasure targeting a primary risk factor — slipping due to slippery underfoot conditions — was considered in a detailed intervention study undertaken by the New Zealand Logging Industry Research Organisation (LIRO) in 1992. The final part of this chapter outlines this study, the lessons learned, and it considers the implications for tackling the current STF injury problem in logging.

9.3 The Effect of Spiked-Soled Boots on Logger Safety, Productivity, and Workload

Analysis of injury data reported to the Forest Industry Accident Reporting Scheme (ARS) maintained by COHFE during the period 1985–1991 indicated that 17.5% of logging lost-time injuries were a result of slips, trips, and falls. These injuries resulted in the loss of 2870 days over the 6-year period. Most (56%) of these STF injuries occurred during the felling and delimbing phase of logging operations, where 37% of the workforce were employed.

In response to the findings of this analysis, LIRO undertook a study to measure the effect of spiked-soled boots under real in-forest conditions (Kirk and Parker 1994). The study investigated the impact of wearing spiked boots on physiological workload, productivity and safety, compared with wearing conventional rubber-soled boots. Spiked-soled logger's boots have approximately 20 1-cm long spikes, similar to golf shoe spikes, screwed into a smooth leather sole.

9.3.1 Method

Four experienced tree fellers were studied during normal work activities. The fellers were working on moderate to steep slopes, felling trees with mean diameters of 48 cm (SD 8.2 cm). A weather station measured air temperatures (i.e., ambient, wet and dry-bulb, and black globe) at 15-minute intervals during the study days, from a position as close as reasonably practicable to the worksite. The activities of the workers were recorded using a continuous time study method. A Husky Hunter™ field computer recorded 24 separate elements of the work cycle. Examples of activities included "walk" (walking between trees), "backcut" (the final saw cut when felling), and "delimbing" (cutting branches off the fallen tree with a chain saw), with the total cycle time being the time taken to fell and delimb the tree. The workers were studied over three days, wearing the conventional boots. Each worker then wore a pair of spiked boots for a 4-week period during which time normal work activities were continued. Following the 4-week period, the continuous time study method was repeated in the same stand and with similar terrain and conditions. In both conditions, the number of times the worker slipped was recorded into the field computer by the observer.

The occurrence of a slip was categorised into three groups according to the location of the worker at that time: working on bare ground, standing on the stem, and standing on slash (i.e., branches, bark, stumps, and foliage on the ground). This way each slip could be related to the ground conditions as well as the specific work activity. A "slip" was recorded if the worker lost their footing, resulting in loss of balance or sliding of the foot.

The heart rate of each faller was recorded at 15-second intervals using a Polar Electro Sport Tester™ PE 3000 heart rate monitor. Task activities were recorded at the same 15-second interval to relate heart rate recordings to the specific activities being undertaken.

9.3.2 Results

Analysis of tree, weather, and terrain data using analysis of variance (ANOVA) indicated that there were no significant differences in these characteristics between the two studies. The delimbing slope and ground slope were, however, significantly steeper ($p < 0.05$) during the second stage of the study when workers were using the spiked boots. This could indicate an expectation of increased slips during the period of spiked boots use, or potential increased workload.

Productivity was not reduced when wearing spiked-soled boots. Analysis of recorded slips indicated a significant ($p < 0.01$) decrease in slips on dirt, slash or the stem when spiked boots were worn (Table 9.8). There was a particularly large reduction in slipping when working on slash.

Average working heart rates were calculated, excluding rest periods or other delays in activities. The recording of climatic conditions allowed comparisons of workloads between the two methods, with no significant differences between wet-bulb globe temperatures. Three out of the four workers reported no significant differences in average heart rates between the two boot types. One worker had a higher heart rate when wearing the spiked boots, which may be explained by the increase slope.

Table 9.8   Location and Frequency of Slips (per 100 Trees Felled and Trimmed)

Location of slip	Bare Ground	Stem	Slash
Rubber-soled boot	43	56	76
Spiked boot	19	13	15

9.3.3 Conclusions

The wearing of spiked-soled forestry boots was associated with a reduction in the rate of slipping for a variety of underfoot conditions (i.e., dirt, slash, logs/stems), in turn having the potential for reducing lost-time injuries caused by STF. Spiked boots are now widely used in the industry.

9.4 Conclusion

A considerable proportion of injuries in the New Zealand logging industry involve STF. Although attempts have been made by forest companies to reduce the risk of STF injuries, including the introduction of spiked-soled boots, up-to-date injury surveillance data has indicated that STF are still a major problem. This failure to significantly reduce STF injury rates over time is evidence for the need for interventions that target more than one aspect of STF risk. It is, perhaps, the relatively low level of perceived injury risk associated with STF that is the greatest barrier to prevention in an industry such as logging. Two further obstacles to STF prevention confound this important factor. Logging work takes place in an uncontrolled, relatively unsupervised, outdoor environment, where natural and man-made slipping and tripping hazards are plentiful. Conventional approaches to STF prevention (e.g., housekeeping, antislip surfaces) are largely inappropriate for this occupational setting. Future prevention efforts will need to look at the potential for risk reduction through measures, such as work organisation, training, improved hazard awareness, and safer working techniques, if real reductions of this costly injury problem are to be achieved.

Acknowledgments

COHFE acknowledges the cooperation of the loggers, contractors, and companies that supplied the data used in this report and that participated in the interviews. The ARS is funded jointly by the Foundation for Research,

Science and Technology, the New Zealand Forest Owners' Association, and the New Zealand Forest Industries Council.

References

Ashby, L., Bentley, T., and Parker, R., 2001, The forest silviculture accident reporting scheme, summary of reports — 2000. COHFE Report, Vol. 2, No. 4. Centre for Human Factors and Ergonomics: Rotorua, New Zealand.

Bates, G., Parker, R., Ashby, L., and Bentley, T., 2001, Fluid intake, hydration status of forest workers — a preliminary investigation. *Journal of Forest Engineering* 12, (2).

Bentley, T. and Haslam, R., 2001, Identification of risk factors and countermeasures for slip, trip and fall accidents during the delivery of mail. *Applied Ergonomics*, 32, 127–134.

Bentley, T.A., Parker, R.J., Ashby, L., Moore, D.J., and Tappin, D.C., 2002, The role of the New Zealand forest industry injury surveillance system in a strategic ergonomics, safety and health research programme. *Applied Ergonomics*, 33, 395–403.

Evanson, T., Parker, R., Ashby, L., and Bentley, T., 2001, Analysis of lost time injuries — 2000 logging (accident reporting scheme statistics). COHFE Report, Vol. 2, No. 6. Centre for Human Factors and Ergonomics: Rotorua, New Zealand.

Kawachi, I., Marshall, S., Cryer, C., Slappendel, C., Laird, I., and Wright, D., 1994, The epidemiological diagnosis of injury to forest workers: the final and amalgamated report. Milestone 6 report for the Forest Industry Research Group to the Accident Compensation Corporation. ACC: Wellington.

Kirk, P. and Parker, R., 1994, The effect of spiked boots on logger safety, productivity and workload. *Applied Ergonomics*, 25, 106–110.

Kirk, P. and Paterson, T., 1996, Increased safety and performance through "smart food." LIRO Report, Vol. 21, No. 26. LIRO: Rotorua, New Zealand.

Lilley, R., Feyer, A.-M., Kirk, P., and, Gander, P., Injury in forest work in NZ: the role of work practices. Personal communication.

Myers, R. and Fosbroke, D.E., 1994, Logging fatalities in the United States by region, cause of death and other factors — 1980 through 1988. *Journal of Safety Research*, 25, 97–105.

New Zealand Forest Owners Association, 2004, *Facts & figures 2004/2005* (New Zealand Forest Owners Association: Wellington).

Parker, R. and Bentley, T., 2000, Skid work injuries 1995–1999, COHFE Report, Vol. 1, No. 4, 2000. Centre for Human Factors and Ergonomics: Rotorua, New Zealand.

Parker, R., Ashby, L., and Evanson, T., 2002, Analysis of lost time injuries — 2001 Logging (Accident Reporting Scheme statistics), COHFE Report, Vol. 3, No. 4. Centre for Human Factors and Ergonomics: Rotorua, New Zealand.

Chapter 10

Falls during Entry/Egress from Vehicles

Fadi Fathallah

CONTENTS

10.1 Introduction

Injuries occurring during entry/egress from commercial vehicles affect many drivers and passengers and translate into a substantial cost to business. The main objectives of this chapter are to give an overview of the problem of vehicle entry/egress-related falls, as well as to summarize the findings of a study that focused on quantifying impact forces and slip potential during various egress methods from five common commercial vehicle configurations. Factors that may contribute to falls during entry/egress from vehicles will also be discussed, along with various approaches that can be adopted to reduce the prevalence of these types of falls.

10.2 Cost and Prevalence

Falls during entry/egress from commercial vehicles have existed since the introduction of motor trucks. Miller (1976) indicated that about one-fourth of all truck driver injuries in the United States are associated with slips and falls in and around the tractor. The author suggested that many truck-related injuries occur during tractor cab entry and egress and believed that these injuries account for the majority of long-term disabilities, especially back injury disabilities (Miller 1976). This problem has seen dramatic increase in its prevalence and related costs; however, due to developments in truck design as well as the introduction of and changes in the U.S. workers' compensation insurance systems (Heglund 1987). Since their inception, trucks generally have increased in size and have been introduced in high-profile designs such as the cab-over-engine. These factors combined, resulted in a substantial increase in the vertical ground-to-cab distance, which exposed drivers and passengers to a higher risk of falling during entry/egress from such vehicles. At the same time, medical costs and workers' compensation benefits have increased substantially, which placed a major economic burden on the trucking industry (Heglund 1987).

Injuries from exiting trucks and trailers represent a substantial cost to U.S. businesses that operate these types of equipment. A study of workers' compensation claims for the period between 1983 and 1986 reported that, of the 16 major injury categories, "Slips/Falls-Elevation" ranked fifth in both number and cost of claims (Heglund 1987). "Highway Vehicles" was one of 49 subcategories within "Slips/Falls-Elevation." This subcategory accounted for 8% of all reported falls and 7% of the total dollar cost to industry. The average cost of falls from highway vehicles was 60% higher than the average cost from all subcategories and 56% higher than the average manual materials handling (MMH) injury claim (Heglund 1987). More recent claims analyses reported that "falls from highway vehicles" still constitutes a major concern in terms of prevalence and cost. The results demonstrated that the average cost of a "fall from highway vehicle" claim between 1993 and 1995 was nearly twice as high as the average claim cost for "all" claims (Fathallah et al. 2000). Furthermore, according to the Bureau of Labor Statistics (BLS), slips and falls are the leading cause of lost workdays to trucking and courier companies (23% of total reported cases in 1997), second only to overexertion cases (28% of total reported cases in 1997). "Falls on the same level" accounted for 11%, "falls from an elevation" accounted for 9%, and "slips or trips with no fall" accounted for the remaining 3% of the total lost workday cases (Bureau of Labor Statistics 1999).

Although most commercial vehicles are equipped with steps and rails to facilitate safe cab egress, many drivers still choose not to use these aids and simply jump off. In a survey of delivery drivers at two companies in the United Kingdom, Nicholson and David (1985) reported that more than half of the respondents admitted to jumping from the cab when leaving. Additionally, based on observations made at a truck stop, Heglund (1987) reported that 30% of drivers who exited high-profile cab-over-engine tractors (cab level 1.45 m high) at least partially jumped out. Although no specific evidence has been reported on slips and falls during entry/egress from the back of commercial equipment, it is fairly common for the backs of trucks and trailers (as compared with tractors) not to have proper entry/egress aids. The backs of many trucks/trailers have no handrails (e.g., some flatbed trucks) and no appropriate steps; thus, a person often has no alternative but to either jump or awkwardly climb up/down. This can be expected to increase the risk of slips and falls during entry/egress from the back of trucks and trailers.

The group at highest risk of falling from commercial vehicles are the drivers, with a varying degree of exposure depending on the nature of their work. The long-haul driver may be expected to have the least frequency of exposure, whereas drivers involved in multi-stop, pick up, and delivery work can be expected to be at increased risk from falling during truck entry and egress (Heglund 1987). Shop, yard, and terminal personnel are also exposed, nonetheless at lower degrees than drivers. These personnel enter and egress cabs while performing maintenance work, moving vehicles in and out of shops, hooking/unhooking combination units, or spotting equipment at loading docks (Heglund 1987).

10.3 Risks of Improper Entry/Egress

Improper entry/egress from commercial vehicles can lead to several potential driver injuries. The driver may be exposed to two major potential injury risks during egress from a commercial vehicle. First, it is expected that whenever a driver jumps/partially jumps out of a vehicle cab or trailer, due to lack of use, poor design, or absence of egress aids, the forces on the body joints will be substantially greater than the forces experienced during situations where egress aids are fully utilized. Although no specific data are available on the risks of jumping from commercial vehicles, the literature indicates that many sports that involve jumping, such as the vertical jump, volleyball, and basketball have a high number of injuries that occur during landing. The most affected areas include the ankles, knees, and the lower back (Dufek and Bates 1991; Ferretti et al. 1992; Kannus and Natri 1997; Maehlum and Daljord 1984; Roberts and Roberts 1996). Parachuting injuries are also dominated by ankle and spinal injuries caused during landing impact (Hallel and Naggan 1975). Furthermore, jumps from the back of military transport trucks have been anecdotally associated with ankle and knee injuries (Rice 1998). The exact injury mechanisms of landing from a jump are not fully understood and depend on many factors such as landing technique, landing height, footwear, and surface type. Nonetheless, it is clear that jumping from an elevation could subject various body joints to excessive levels of force during the landing phase. This may, in turn, lead to various acute or chronic musculoskeletal injuries.

The second risk of injury involves the potential for the driver to slip and fall immediately after landing. This risk depends on many factors such as the material of the landing surface, the presence of a surface contaminant (e.g., oil), the type of shoes worn by the driver, and other factors. It is expected, however, that the potential for slipping and falling after jumping from commercial vehicles exposes the driver to the risk of severe injury to various body parts and could potentially lead to a fatal injury.

Little quantitative data are available to demonstrate the two major potential risks associated with improper egress techniques, as discussed previously. The following section summarizes the results of a study that addressed landing forces and their potential effects on the body, as well as slip potential during the landing phase. The details of the study are presented elsewhere (Fathallah and Cotnam 2000; Fathallah et al. 2000).

10.4 Impact Forces and Slip Potential during Vehicle Egress

The purpose of this study was to quantify impact forces and "slip potential" with icy conditions (wet and dry ice). Data were collected for alternative methods of exiting various commercial vehicle configurations, commonly found in the trucking industry.

10.4.1 Methods

Ten healthy male subjects participated in the study; they were screened to ensure the absence of any current or previous musculoskeletal disorders. The study considered five vehicle configurations:

1. A cab-over-engine (COE) tractor
2. A conventional tractor
3. A delivery step-van
4. The back of an 8-m "box" trailer
5. The back of a 4.3-m "cube" van (Figure 10.1)

Table 10.1 describes the experimental conditions investigated. The COE had two grab-rails along the side of the cab and three steps. The conventional tractor had one grab-rail and two steps, whereas the step-van had three steps and a grab-rail along the exit door. The subjects were instructed to use all three steps when leaving the van (last step at 0.43 m). In certain conditions with the delivery step-van, the subjects carried a $0.23 \times 0.28 \times 0.30$ m 9.5-kg box. The trailer had two custom-installed steps, and the cube-van had a step that ran along the whole of the rear of the van. The trailer had an edge that ran along its sides, which was used as a grab-rail, whereas the cube-van had a 0.89-m grab-rail that was custom installed on its right side.

A three-dimensional force plate (Bertec 4060A; Bertec, Worthington, Ohio) was used to capture the impact forces in each of the experimental conditions (see Figure 10.1). For both tractors, the cab-level conditions required the subject to "squat" jump from the cab level facing forward (away from the tractor), whereas all other conditions (using steps and rails) required the subjects to face the tractor. For the step-van conditions, the subjects were asked to exit facing away from the vehicle. For the trailer-level exits within the back of trailer conditions, the subject was asked to squat jump either forward or backward while supporting his body by placing one hand on the trailer's surface. For exits that required using the steps, the subjects faced the back of the trailer while using part of the edge of the trailer as a grab-rail and the appropriate step(s) (step 1 or step 1 and 2).

The vector (geometric) sum of all three forces (lateral [x], anterior-posterior [y], and vertical [z]) was calculated (square-root[$x^2 + y^2 + z^2$]). This figure was then divided by the subject's body weight to represent the total impact force magnitude in multiples of body weight observed in a given trial. For each trial, the "required coefficient of friction" (RCOF) was determined as the ratio of peak of the vector (geometric) sum of the horizontal forces (x and y), over peak vertical (z) force (square-root[$x^2 +$

(a) COE tractor

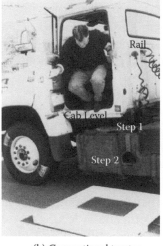

(b) Conventional tractor

(c) Step-van

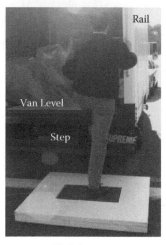

(d) Cube-van

(e) Box trailer

Figure 10.1   Five vehicle configurations investigated in the study of impact forces and slip potential during entry/egress: (a) COE tractor; (b) conventional tractor; (c) step-van; (d) cube-van; and (e) box trailer.

y^2]/z) (Hanson et al. 1999). A given situation is considered "unsafe" and may lead to a slip or fall whenever the RCOF of the landing surface exceeds the dynamic environmental or "available COF" (ACOF) for a particular surface condition (Hanson et al. 1999). Thus, in this study, the difference between the ACOF and the RCOF (ACOF − RCOF) indicates the potential for slipping, or slipping and falling on icy surfaces (dry and

Table 10.1 Experimental Conditions for Each Vehicle Configuration

Condition	Condition Number	Exit Height (m)	MBW Average (SD)	RCOF Average (SD)	Probability of a "Fall" Wet/Dry Ice	POST HOC[a]
Cab Exit: COE Tractor						
Squat jump forward from cab level	1	1.25	7.04 (1.57)	0.26 (0.06)	0.81/0.32	3, 4
Exit using rail and first step	2	0.97	6.45 (1.58)	0.28 (0.08)	0.82/0.39	3, 4
Exit using rail, first, and second steps	3	0.69	2.13 (1.11)	0.18 (0.5)	0.61/0.15	4
Exit using rail, first, second, and third steps	4	0.38	1.43 (0.50)	0.13 (0.06)	0.46/0.09	—
Cab Exit: Conventional Tractor						
Squat jump forward from cab level	1	1.07	7.21 (1.83)	0.29 (0.11)	0.84/0.41	3[b]
Exit using rail and first step	2	0.86	5.07 (1.47)	0.29 (0.12)	0.82/0.41	3[b]
Exit using rail, first, and second steps	3	0.43	1.77 (0.51)	0.22 (0.05)	0.72/0.23	—
Cab Exit: Step Van						
Normal exit with no rail	1	0.43	2.69 (0.76)	0.17 (0.04)	0.60/0.14	—
Normal exit with rail	2	0.43	2.06 (0.66)	0.19 (0.04)	0.64/0.16	—
Fast exit with no rail	3	0.43	3.06 (0.66)	0.21 (0.06)	0.70/0.23	—
Normal exit with no rail and carrying a package	4	0.43	2.92 (0.86)	0.18 (0.06)	0.61/0.17	—
Normal exit with rail and carrying a package	5	0.43	1.93 (0.65)	0.21 (0.06)	0.68/0.20	—
Fast exit with no rail and carrying a package	6	0.43	3.50 (0.66)	0.19 (0.05)	0.64/0.17	—

Table 10.1 Experimental Conditions for Each Vehicle Configuration (Continued)

Condition	Condition Number	Exit Height (m)	MBW Average (SD)	RCOF Average (SD)	Probability of a "Fall" Wet/Dry Ice	POST HOC[a]
Back Exit: Trailer						
Squat jump-forward	1	1.14	6.44 (2.42)	0.21 (0.03)	0.72/0.2	2, 4
Squat jump-backward	2	1.14	5.92 (1.40)	0.33 (0.05)	0.92/0.57	3, 4
Exit using rail and first step	3	0.76	2.57 (1.01)	0.21 (0.05)	0.71/0.22	4
Exit using rail and first and second step	4	0.51	2.11 (0.69)	0.15 (0.04)	0.54/0.11	—
Back Exit: Cube Van						
Squat jump-forward	1	0.71	5.48 (1.76)	0.23 (0.06)	0.75/0.27	3
Exit using rail from van level	2	0.71	1.92 (0.66)	0.22 (0.06)	0.71/0.24	3
Exit using rail and step	3	0.36	1.42 (0.29)	0.14 (0.04)	0.49/0.09	—

Note: Multiples of body weight (MBW) and required coefficient of friction (RCOF) descriptive statistics, probability of a "fall" occurring for *assumed* wet and dry ice conditions, and post hoc analyses are also listed; SD = standard deviation.

[a] Condition number(s) that is (are) significantly different from a corresponding condition listed in the second column ($p < 0.05$).

[b] $p < 0.1$.

wet). Furthermore, the probability of a "slip or fall," and the probability for a "fall" were quantified for the two icy conditions based on the two logistic regression models reported in the literature (Hanson et al. 1999).

10.4.2 Study Results and Implications

Table 10.1 presents basic descriptive statistics on the RCOF, multiples of body weight (MBW), as well as the probability that a "fall" might occur on landing with icy conditions.

As mentioned earlier, many truck drivers do not make use of the egress aids provided in commercial tractors and choose to simply jump or partially jump from the cabs of their vehicles. These unsafe acts may increase the risk of musculoskeletal injuries due to the impact forces when landing or the risk of slipping and falling immediately after landing (Grönqvist 1999).

For the tractor cab egress conditions, the results demonstrated that the lowest impact forces were observed when the subjects used both the grab-rail(s) in combination with the entry/egress steps. On the other hand, impact forces reached an average of over seven times body weight and as high as 12 times body weight when the subjects jumped from the (COE) cab-level (5800 N average and 11,000 N maximum approximately). It is difficult to determine how much of this force is transmitted through each joint of the body; however, it is expected that the ankle joint will be subjected to the majority of the force, given its proximity to the contact surface (force plate). Limited information is available about the fracture tolerance of the ankle joint. Yoganandan et al. (1997) demonstrated that fractures at the proximal tibial end of the ankle and to the plantar side of the ankle occurred in cadaveric specimen when subjected to mean maximum forces of 10.2 kN (1.5 SD) and 15.1 (2.7 SD), respectively. Klopp et al. (1997) have also demonstrated that the 50% probability of injury to the plantar side of the ankle could occur at a 9.3-kN contact force. These findings indicate that forces observed during the jump conditions in this study approached the fracture tolerance of the ankle joint. In addition, these forces are near the compressive strength of the L5/S1 spinal joint (Adams and Dolan 1995; Jäger and Luttmann 1991). Given that the structure of the human body cannot be considered rigid; however, the ground impact reaction force is expected, in most cases, to be attenuated for higher-level joints such as the knees and the lower back. The level of attenuation (or in some instances magnification) depends on many factors such as the natural frequency of the joints, joint posture, and acceleration (Pope et al. 1997). Note though, that even with 30–50% attenuation of ground reaction force, the forces experienced at the spine level would still be considered substantial and may exceed some recommended exposure limits (e.g., the NIOSH limit of 3.4 kN) (Waters et al. 1993).

The results have demonstrated clearly that with a slippery, contaminated surface (wet or dry ice), the potential for slipping (and falling) is expected to increase greatly when egress aids are not used. The probability of a fall occurring almost reached certainty under conditions where jumps from the cab level or back of trailer were performed. The increased slip potential during "squat" jumps is also related to the squeeze-film pressure effect due to large impact forces (see Chapter 2); however, this was not examined in the present study. (The available friction values were determined using force levels during simulated normal walking [Grönqvist and Hirvonen 1995].) Note that for both the COE and conventional tractors, the "jump from step 1" was not significantly different from the "squat jump from cab level." This emphasizes the need to fully utilize egress aid systems, especially under slippery conditions such as ice, or other surface contamination. Only when the egress system was fully utilized did the risk of a fall decrease to an acceptable level.

For the delivery step-van, using the handrail significantly reduced the impact force for both the "with package" and "without package" conditions (a reduction of over 40%). This decrease may result in a substantial reduction in the potential "cumulative" effect of egress impact forces. This is especially important in the case of the delivery van driver who is expected to leave the vehicle many times throughout the day (as compared with a tractor-trailer truck driver).

The conditions resulting from exiting from trailers and the back of trucks have demonstrated the importance of both installing and using egress aids. For the box trailer, squat jumping (forward and backward) from the trailer level resulted in high impact forces (as high as 11 times body weight) and increased levels of RCOF. These forces and RCOFs were, on average, two to three times higher than the forces and RCOFs observed when fully utilizing the retrofitted step system (i.e., a two-thirds reduction in impact force when using the system). Similar reduction was observed with the cube van when comparing the squat jump condition to the "exit from step-level" condition. These reductions would be expected to reduce the risk of injuries (cumulative or acute) among people who perform tasks such as loading/unloading from trailers and the back of trucks on a frequent basis. It should be noted that many trailers and flatbed trucks do not offer adequate aids to facilitate safe entry or egress. In many instances, entry/egress aids such as ladders or rails are optional features to these units instead of standard equipment. Thus, proper operator training alone may not be sufficient to contain the risks associated with exiting commercial vehicles if these units are not equipped with the adequate entry/egress aids.

It should be emphasized that the jumping conditions (e.g., squat jump from cab level) investigated in this study were collected under ideal situations in terms of speed of egress, the manner of egress, and surface quality of the steps and landing area. The subjects were given ample time to prepare for the conditions and were carefully trained in executing the exits. Whereas, in real-world delivery situations, where the driver may jump more hurriedly out of the cab, the speed of egress and the manner in which the jump is executed may cause the impact forces and the RCOF to be even higher than those reported in this study. A combination of these potentially high forces and required coefficient of friction values during unaided egress (e.g., a jump), with less than ideal landing surface conditions (e.g., icy or wet surface), can be expected to increase the likelihood of slipping and falling on landing.

In summary, this study clearly demonstrated the potential benefits of using egress aids, such as steps and grab-rails in commercial vehicles, and highlighted the importance of installing these aids in units that do not have them such as trailers. Among the reasons that drivers do not use the aids, besides saving time, are design issues such as anthropometric mismatch (e.g., grab-handle is difficult to reach), size and location of aids, and so forth. Therefore, an approach that emphasizes optimal design of entry/egress aids, coupled with driver training in proper use of aids and education on the potential risks associated with unsafe entry/egress techniques (e.g., jumping), is desirable to minimize vehicle-related falls and musculoskeletal injuries. The findings and conclusions of this experimental study are consistent with the results and recommendations of a recent study that quantified forces on the lower limbs when stepping down from fire fighter vehicles (Giguere and Marchand 2005).

10.5 Approaches to Preventing Vehicle-Related Falls

Heglund (1987) described four basic approaches to preventing falls around commercial vehicles:

1. Specifying access systems in equipment purchase
2. Modifications to existing equipment
3. Maintenance of access systems
4. Training and supervision

The discussion in this chapter continues to consider additional fall-related hazards found in truck/trailer docking areas, concluding with a brief consideration of work organization issues.

10.5.1 Newly Purchased Equipment

Although retrofitting an access system into existing vehicles might be a viable control, it is better to equip newly purchased vehicles with well-

Figure 10.2   Example of the three-point entry/egress system.

designed access systems from the outset. New systems should have the following characteristics:

1. Allow the use of the "three-point system" (i.e., minimum of three out of the four extremities, the two hands and two feet, should be coupled to the equipment at all times, Figure 10.2)
2. Provide a systematic entry/egress approach, where the user automatically starts to enter or egress with the correct leg
3. Designed in a manner such that when the user locates one of the access components, the other components are easily and naturally located during entry/egress
4. Have self-cleaning, slip-resistant step surfaces to improve shoe grip and reduce long-term maintenance
5. Take into consideration the anthropometric characteristics of the driver population

The issue of trailer access requires particular mention. Many trailers do not offer adequate systems to facilitate safe entry/egress. This situation may in part be due to a lack of standards concerning access requirement for the rear of trailers. Access system components such as ladders or rails are frequently optional features on trailers instead of standard equipment from the manufacturer. This may expose trucking personnel to serious potential for slip and fall injuries, especially for those who enter/egress trailers without the use of loading docks. The horizontal under-ride guard on trailers (intended to prevent cars from going under a trailer where a collision occurs, Figure 10.1e) is commonly the only intermediate step between the ground and the trailer deck; however the horizontal bar is typically 64–76 cm from the ground, above the knee height for more than 95% of the adult male population (Pheasant 1998).

It is important that driver population anthropometric characteristics are considered in access system evaluation and design. The population of professional drivers and material movers has changed over the last 20 years. Between 1975 and 1995, there has been a 25% increase in women employed in transportation and material moving in the U.S. Women, who are generally shorter in stature than their male counterparts, therefore need to be considered in the design of vehicle access systems. The standards of the U.S. Federal Motor Carrier Safety Administration (FMCSA) indicate that any person entering or exiting the cab or accessing the rear portion of a high profile COE truck tractor shall be afforded sufficient steps and handholds/grab-rails, or deck plates to allow the user to have at least three limbs in contact with the truck or truck tractor at any time (FMCSA 2002). FMCSA standards and recommendations by the Technical and Maintenance Council of the American Trucking Associations specify that the vertical height of the first step shall be no more than 61 cm from the ground level; however (American Trucking Associations 2001; FMCSA 2002). This design standard is above the knee height of the majority of the population. These and other vehicle access systems should be evaluated and designed to accommodate extremes of the targeted user population, to ensure effective utilization and reduce the potential for fall-related injuries.

10.5.2 *Modifying Existing Equipment*

Modifications to existing vehicles can be performed to reduce exposure to the risk of falling around commercial vehicles. Given the wide variety of vehicles, it is difficult to recommend one modification that would fit all. The unique features of each vehicle should be taken into consideration. Examples of possible modifications include (Fathallah and Cotnam 2000; Heglund 1987):

1. Repositioning of handholds/grab-rails or installing longer ones
2. Use of slip-resistant material on step surfaces
3. Use of self-cleaning, slip-resistant metal step surfaces
4. Replacing an entire step system
5. Installing a new system (e.g., back of trailers)

10.5.3 Access System Maintenance

The issue of maintenance is an important one and is often ignored. Even the best access systems can be degraded if damaged or neglected. Unfortunately, it is not common to have the access system as part of the vehicle preventive maintenance programmes (Heglund 1987). All system components should be inspected periodically, especially slip-resistant material, as part of the vehicle scheduled maintenance checks.

10.5.4 Training and Supervision

Training and supervision are important aspects of an effective control strategy. The manner in which the driver enters or egresses from a vehicle is as important as the design of the access system. A well-designed system that is improperly used could be as ineffective as a poorly designed system. An effective training programme should be part of the overall fleet safety programme and should be designed around the following elements of safe entry/egress (Heglund 1987):

> **Wear proper footwear.** Good sturdy footwear with slip-resistant soles will provide maximum coupling with step surfaces.
>
> **Know your equipment.** Vehicles and their access systems vary. Know where step ladders, grab bars, or handholds are located.
>
> **Use the three-point system.** Have at least three extremities (either two hands and one foot, or two feet and one hand) in contact with the vehicle or the ground at all times. This system allows a person to have maximum stability and support while entering/egressing commercial vehicles and, therefore, reducing the likelihood for slipping and falling.
>
> **Look before exiting.** Be aware of the ground or pavement before exiting. Watch for holes, uneven pavement, debris, ice, snow, or anything else that may affect the footing.
>
> **Exit in the right direction.** Climbing out of the cab should be in the same direction as entering it. If the driver faced the vehicle when entering, he or she should face the vehicle when exiting. This optimizes the use of the handholds/grab-rails, allows full contact between the foot and the steps, and helps maintain body balance.
>
> **Keep your hands free.** Drivers should not carry anything while entering or exiting. Hands should be free to grip handholds or grab-rails.

Companies can use a combination of approaches to provide an effective training programme, such as the use of videos, brochures, computer-based training (CBT), e-mails, and bulletin boards. Well-rounded and properly supervised training arrangements coupled with well-designed access systems should provide an effective approach for controlling falls-related injuries in the trucking industry.

10.5.5 Docking Areas

The docking areas of trucks and trailers present several types of hazards to workers when entering or exiting vehicles. Dock accidents can range from a minor slip on a dock plate to more serious, possibly fatal injuries, in the case of a fall off the dock edge. Key issues to consider around docking areas include:

1. **Use of Dock Levellers or Plates.** Dock levellers are either mechanical (raised and lowered by hand) or hydraulic (powered) systems (see Figure 10.3) that allow the levelling between the trailer/truck bed and the docking area. Although, levellers are typically installed during initial construction of shipping and receiving areas, retrofitting them into existing docking areas is recommended due to their added safety benefits. Dock plates or dockboards (Figure 10.3) are commonly found in older facilities. They are not as safe as levellers because they must be carried, slid, positioned, and adjusted manually, thus they would be recommended only where it really is not feasible to install dock levellers.

2. **Maintenance and Cleaning.** The equipment used around the docking area (e.g., dock levellers) should be well maintained to assure proper and efficient use. In addition, due to commonly high worker, equipment, and material traffic around docking areas, the surfaces of these areas tend to be exposed to floor contaminants such as oils, mud, and water. Frequent and scheduled cleaning of these surfaces areas minimizes the likelihood of slip and fall-related injuries.

3. **Enclosed Bays.** Equipping the loading bays with enclosures (partial or complete) will help avoid weather-related slip/fall hazards such as rain or snow.

10.5.6 Work Organization

Efficient organization of the work performed by commercial vehicle drivers and assistants should include the reduction of the frequency of entry/egress required in a typical work shift. This could be accomplished

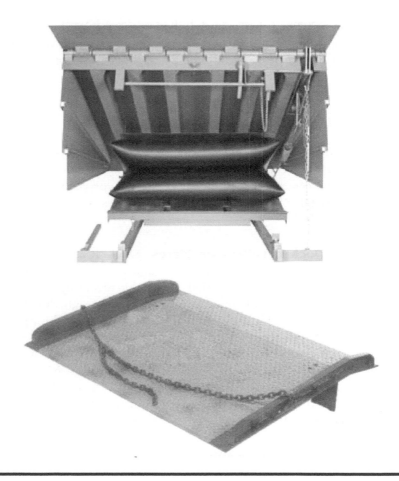

Figure 10.3 An example of a dock leveller (top) and a dock plate (bottom).

by conducting a detailed job/task analysis. The analysis should help identify tasks or subtasks where entry/egress from the vehicle can be reduced through task consolidation, elimination, or re-assignment.

10.6 Summary

Personnel who operate or work around commercial vans, trucks, and trailers are at risk of serious fall-related injuries during entry/egress from these vehicles. This chapter has provided an overview of the problem of vehicle entry/egress falls and has summarized the findings of a study that focused on quantifying impact forces and slip potential, during various methods of egress, from five common commercial vehicle configurations. The chapter has discussed factors that may contribute to falls during entry/egress along with various approaches that can be adopted to reduce the prevalence of vehicle-related falls.

Acknowledgment

The experimental study presented in this chapter was funded and performed by the Liberty Mutual Research Center for Safety and Health, Hopkinton, Massachusetts.

References

Adams, M.A. and Dolan, P., 1995, Recent advances in lumbar spinal mechanics and their clinical significance. *Clinical Biomechanics*, 10, 3–19.

American Trucking Associations, 2001, *Recommended practices manual: truck and truck tractor access systems (RP404B-1)*. Alexandria, VA: Technology and Maintenance Council of the American Trucking Associations.

Bureau of Labor Statistics, 1999, Lost-worktime injuries and illnesses: characteristics and resulting time away from work, 1997 (News Release USDL 99-102). Washington, D.C.: United States Department of Labor.

Dufek, J.S. and Bates, B.T., 1991, Biomechanical factors associated with injury during landing in jump sports. *Sports Medicine*, 12, 326–337.

Fathallah, F.A. and Cotnam, J.P., 2000, Maximum forces sustained during various methods of exiting commercial tractors, trailers and trucks. *Applied Ergonomics*, 31, 25–33.

Fathallah, F.A., Grönqvist, R., and Cotnam, J.P., 2000, Estimated slip potential on icy surfaces during various methods of exiting commercial tractors, trailers, and trucks. *Safety Science*, 36, 69–81.

Ferretti, A., Papandrea, P., Conteduca, F., and Mariani, P.P., 1992, Knee ligament injuries in volleyball players. *American Journal of Sports Medicine*, 20, 203–207.

Federal Motor Carrier Safety Administration (FMCSA), 2002, Step, handhold, and deck requirements for commercial motor vehicles (Employee Safety and Health Standards, Subpart L — Section 399.201–399.211). Washington, D.C.: U.S. FMCSA.

Giguere, D. and Marchand, D., 2005, Perceived safety and biomechanical stress to the lower limbs when stepping down from fire fighting vehicles. *Applied Ergonomics*, 36, 107–19.

Grönqvist, R., 1999, Slips and falls. In S. Kumar (Ed.), *Biomechanics in ergonomics* (London: Taylor and Francis), pp. 351–375

Grönqvist, R. and Hirvonen, M., 1995, Slipperiness of footwear and mechanisms of walking friction on icy surfaces. *International Journal of Industrial Ergonomics*, 16, 191–200.

Hallel, T. and Naggan, L., 1975, Parachuting injuries: a retrospective study of 83,718 jumps. *Journal of Trauma*, 15, 14–19.

Hanson, J.P., Redfern, M.S., and Mazumdar, M., 1999, Predicting slips and falls considering required and available friction. *Ergonomics*, 42, 1619–1633.

Heglund, R.E., 1987, SAE Technical paper series: falls entering and exiting heavy truck cabs. Truck and Bus Meeting and Exposition, Dearborn, Michigan. Warrendale, Pennsylvaia: Society of Automotive Engineers (SAE).

Jäger, M. and Luttmann, A., 1991, Compression strength of lumbar spine elements related to age, gender, and other influencing factors. In P.A. Anderson, D.J. Hobart, and J.V. Danoff (Eds.), *Electromyographical kinesiology* (Elsevier Science Publishers B.V. [Biomedical Division]), pp. 291–294.

Kannus, P. and Natri, A., 1997, Etiology and pathophysiology of tendon ruptures in sports. *Scandinavian Journal of Medical Science and Sports*, 7, 107–112.

Klopp, G., Crandall, J., Hall, G., Pilkey, W., Hurwitz, S., and Kuppa, S., 1997, Mechanisms of injury and injury criteria for the human foot and ankle in dynamic axial impacts to the foot. *1997 International IRCOBI Conference on the Biomechanics of Impact* (Bron, France: IRCOBI).

Maehlum, S. and Daljord, O.A., 1984, Acute sports injuries in Oslo: a one-year study. *British Journal of Sports Medicine*, 18, 181–185.

Miller, J.M., 1976, Efforts to reduce truck and bus operator hazards. *Human Factors*, 18, 533–550.

Nicholson, A.S. and David, G.C., 1985, Slipping, tripping and falling accidents to delivery drivers. *Ergonomics*, 28, 977–991.

Pheasant, S., 1998, *Bodyspace:anthropometry, ergonomics, and the design of work* (2nd ed.). (London; Bristol, Pennsylvania: Taylor & Francis).

Pope, D.P., Croft, P.R., Pritchard, C.M., Silman, A.J., and Macfarlane, G.J., 1997, Occupational factors related to shoulder pain and disability. *Occupational and Environmental Medicine*, 54, 316–321.

Rice, V.J., 1998, Personal communication: Colonel Valerie J. Rice, U.S. Army Medical Department Center and School. Fort Sam Houston, Texas.

Roberts, S.N. and Roberts, P.M., 1996, Tournament water skiing trauma. *British Journal of Sports Medicine*, 30, 90–93.

Waters, T.R., Putz-Anderson, V., Garg, A., and Fine, L.J., 1993, Revised NIOSH equation for the design and evaluation of manual lifting tasks. *Ergonomics*, 36, 749–776.

Yoganandan, N., Pintar, F.A., Kumaresan, S., and Boynton, M., 1997, Axial impact biomechanics of the human foot–ankle complex. *Journal of Biomechanical Engineering*, 119, 433–437.

Chapter 11

Behaviour and the Safety of Older People on Stairs

Roger Haslam, Denise Hill, Joanne Sloane,
Peter Howarth, and Katherine Brooke Wavell

CONTENTS

11.1 Introduction

As described in Chapter 5, older people have an increasing vulnerability to falling, to the extent that 1 in 3 adults over age 65, and nearly half of those over 80, fall each year (Prudham and Evans 1981). Approximately half of all recorded fall episodes occurring among independent community dwelling older people happen in their homes and immediate home environments (Lord et al. 2001). Estimates for the United Kingdom, based on the Home Accident Surveillance System (HASS), indicated that 373,000 older people in the United Kingdom received injuries from a fall in the home in 2002, severe enough to require attendance at a hospital accident and emergency department (DTI 2003). Within these figures, older people falling on stairs in their homes comprise a significant category, resulting in excess of 500 deaths and 51,000 hospital casualty department attendances each year. These statistics do not include patients seeing their general practitioner or those not seeking treatment. A similar pattern exists in the United States (Startzell 2000). Worldwide, it has been argued that falls on stairs rival road accidents as a leading cause of accidental injury and death (Pauls 1991; Templer 1992).

Although researchers generally agree that older users are at increased risk of falling on stairs than younger adults, the extent of this is difficult to quantify due to the absence of data on stair usage by different age groups (Templer 1992). When older individuals do fall on stairs, their injuries tend to be more serious, with fractures more common (Pauls 1985; Nagata 1993). The cost to health services of treating these patients is substantial. Falls also have serious psychological and social consequences for older people, affecting mobility, confidence, and quality of life (Delbaere et al. 2004).

Personal factors contributing to fall accidents among older people are well known and include decreased balance ability, disturbed gait, cognitive impairment, reduced strength and vision, illness, and side effects from use of medication (e.g., Bath and Morgan 1999; Startzell et al. 2000; Lord et al. 2001; Masud and Morris 2001; Tinetti et al. 1988). In the past, environmental aspects have been estimated as a primary cause in fall accidents in the older population in around one-third of cases (Smith 1990), although more recent studies have not been able to isolate these as a major risk factor independent of other causes (Lord et al. Chapter 5, this volume). Environmental features identified by stair accident investigations for falls involving adults (of all ages) include poor condition of stair surface, objects on stairs, risers too high or too low, narrow goings, absent or poorly designed handrails, and poor lighting (Templer 1992; Roys, Chapter 3, this volume).

Dowswell et al. (1999) suggested that falls among the "young elderly" (65–74) are more likely to involve environmental factors, whereas personal factors are more important in the "oldest" age group (those age 85 and over). This notion is compatible with the model described by Lord et al. (2001), where the risk of falling varies according to the interaction between an individual's physical ability and the demands of the environment. Those with declining physical abilities will be affected increasingly by environmental challenges.

Behaviour, especially when coupled with inadequacies of the surroundings or personal frailty, has been suggested as playing a large part in many falls on stairs (Templer 1992). And it appears reasonable to expect that behaviour will be a mediating influence in the interaction between individual abilities and environmental challenges (Connell and Wolf 1997). In the case of stairs, this might involve either the manner of stair use, or actions that affect the stair environment. For example, it has been suggested that older people may be less able to maintain their stairs in good repair (Healy and Yarrow 1998).

In view of the likely influence of behaviour on risk of falling, it is surprising that there has been only limited research concerning the contribution of this to falls among the older age group (Askham et al. 1990; Connell and Wolf 1997). The study reported in this chapter addressed this problem, investigating how older people use their stairs, why they use them the way that they do, and circumstances likely to affect risk of falling.

11.2 Preliminary Focus Groups

An initial qualitative investigation, reported elsewhere (Haslam et al. 2001), used focus groups to examine older people's knowledge of safety on stairs. It was apparent from the focus groups that use of stairs does become more difficult with increasing frailty, leading to avoidance in some cases. The location of essential facilities in some homes (e.g., the lavatory) may necessitate increased use of stairs for certain individuals. The focus group participants offered intuitive observations of situations and circumstances where risk of falling on stairs might be increased, such as leaving objects on stairs, hurrying on stairs, using stairs in the dark, and carrying items, although they reported that they continued to engage in these despite concerns about safety. In this respect, safety implications from features of the environment appeared to be more readily identified than risks from behaviour. Cleaning stairs can present problems due to difficulties with access or a need to use heavy and awkward equipment. The focus group participants generally recognised that medication and use of alcohol may increase risk of falling, but individuals might not appreciate fully when they personally are at increased risk. Discussion within the groups described situations in which bifocal and varifocal spectacles may affect vision and balance on stairs (see also Chapter 4). Most of the study participants indicated that they had given only limited thought to stair safety before the focus groups.

11.3 Home Interview Survey

Building on the focus groups, interviews were conducted with 157 older people living in their own homes, to collect detailed quantitative information on stair use and factors influencing this. Participants were recruited on a convenience basis from the counties of Nottinghamshire and Leicestershire, using a combination of direct and indirect contact. Sampling was on a quota basis, according to age and gender, using estimated population figures for the United Kingdom (ONS 1999a). Likewise, house type was a sampling dimension, based on national estimates of housing stock with respect to age and type of dwelling (ONS 1999b). Socioeconomic class was assessed using postcode analysis (against data provided by Experian 1999).

At the beginning of each interview, participants received an explanation of the purpose of the study and what the interview would entail. The interviews followed the format given in Table 11.1, with each interview lasting approximately 2 hours. Loughborough University Ethical Advisory Committee approval was obtained before commencing the research. Statistical analysis used chi-squared tests for cross-tabulation relationships and Pearson and Spearman correlation coefficients for assessing interval/ordinal data associations, as appropriate. Qualitative data interpretation was also used in conjunction with quantitative analysis.

11.4 Home Interview Survey Results

A total of 157 participants were interviewed for the survey. The majority of these took place on an individual basis, although interviews with 14 married couples were undertaken with both partners present. Consequently, 150 households were visited in total.

11.4.1 Participant and Housing Details

Details of participants are presented in Table 11.2. To assess the effects of age, the participants were split into three age groups: "youngest" (65–74), "middle" (75–84) and "oldest" (85 and over). In terms of data on physical capabilities, as might be expected, Modified "Modified Barthel Index" (MMBI) scores indicated reduced functional abilities in the oldest

Table 11.1 Interview Schedule

Questionnaire	Content	Basis for Inclusion	Developed by
Chances of falling on the stairs (self-completion questionnaire)	Participants' knowledge of factors affecting stair safety	Assessment of relationship between knowledge and behaviour	Loughborough University
Your stairs (self-completion questionnaire)	Participants' opinions on the design, condition, and safe use of their stairs	Individuals' ability to assess condition. Comparison of results with findings from similar BRE survey	Building Research Establishment
Stair environment (survey undertaken by interviewer)	Design and condition of stairs. Design, colour, and state of stair coverings. Colour and design of wall coverings. Number of handrails and their condition. Objects/items on and around stairs. Lighting provision during the day and night. Position of windows	Objective assessment of stair environment. Comparison with participants' own assessment	Loughborough University
Participant information (structured interview)	Gender. Date of birth. Conditions affecting vision. General medical conditions. Prescribed and non-prescribed medications	Basic information	Loughborough University

Table 11.1   Interview Schedule (Continued)

Questionnaire	Content	Basis for Inclusion	Developed by
Ability to perform daily activities (structured interview)	Participants' ability to perform a range of daily activities	Measure of self-assessed functional ability	Modified from Barthel Index
Stairs in the home (semistructured interview)	Patterns of use of stairs, handrails and lighting Behaviours associated with stair use (e.g., carrying, hurrying, cleaning) Placing objects on stairs Other issues regarding stair use including: footwear, alcohol consumption, behaviour of pets, and receipt of previous stair safety advice	Information on behaviour	Loughborough University
Falls on the stairs (semistructured interview)	History of falls since age 65	Examination of differences in behaviour and attitudes between "fallers" and "non-fallers"	Loughborough University
Physical and physiological measurements	Height, weight, length of feet Functional ability ("grip strength", "rise from stool") Vision (acuity, stereopsis)	Basic information on participants and their functional ability	Loughborough University

Note: BRE = Building Research Establishment.

Table 11.2   Participant Characteristics

	Population Figures for U.K. (%)	Participants in Study (%)
Male	41	28
Female	59	72
Living alone		58
Living with others		42
Social Class		
Groups A and B		40
Groups C1 & C2		28
Groups D & E		32
Age Group		
65–74	54	47
75–84	35	41
85+	11	12
Age Summary		
Mean 77 years	Standard deviation 7.1	Range 65–96

age group (age 85 and over). There was no difference in MMBI score with gender. The "oldest" participants also experienced more difficulty with the "rise from stool" test (p < 0.001). The grip strength of the youngest age group in the study (the only age group for which comparison data are available) was slightly lower than recorded by the Allied Dunbar National Fitness Survey (1992). Allied Dunbar reported mean values of 360 N for men and 220 N for women age 65–74, vs. 349 N and 214 N, respectively, for our participants in this age range.

Details of housing characteristics and stair handrail provision are given in Table 11.3 and Table 11.4. A pitch of 42° is the maximum steepness permitted for private stairs by the Building Regulations in the United Kingdom. All the houses built within the last 30 years had a stair pitch less than 42°, with older houses tending to have stairs steeper than 42° (p < 0.001). Among the households with two handrails, the second had been fitted on the advice of a community occupational therapist or social services advisor on 28% of occasions. For 37% of participants, the suggestion originated from family or friends, with the remainder (35%) stating that it was their own idea. Although many participants with a single handrail fitted to their stairs said they thought they would benefit from another handrail in the future (23%), this does not necessarily mean they

Table 11.3   Housing Characteristics (n = 150)

	House Types in U.K. (%)	House Types in Sample (%)
Detached	21	37
Semidetached	32	27
Terraced	28	20
Purpose-built apartment	13	13
Other types of houses	6	3
Households with stair pitch > 42°	—	65
Households with only a single toilet, upstairs	—	33
Households with only a single toilet, downstairs	—	8
Households with toilets upstairs and downstairs	—	59
Age of House	Median 41 years	Range 6–400 years

Table 11.4   Handrail Provision

Households with one handrail	59%
Households with two handrails	34%
Flights where handrail in need of repair	10%
Of those households with two handrails, second rail fitted by the householder	70%

will proceed to obtain one. Participants cited various problems as deterring them from obtaining a second rail, including the expense and knowing whom to approach to assist with installation.

In 41% of households, windows were positioned part way over the landing or completely over stairs. Where windows were in these positions, a number of participants reported difficulty accessing the window to clean it (27%), open it (20%), or to change curtains (26%). Problems arise due to both the position of windows and the functional ability of the individual. The older the age group the more likely it was that participants had problems cleaning the landing window ($p < 0.01$), opening the window ($p < 0.05$) and changing the curtains at the window ($p < 0.001$).

11.4.2 Use of Stairs

When asked about their use of stairs, 32% of interviewees reported that they avoid using stairs wherever possible, with the "oldest" age group using stairs less frequently than the "youngest" and the "middle" age groups (p < 0.01). No differences were reported in the frequency of stair use for "fallers" and "non-fallers" in the sample or between those living with others or living alone. Most interviewees (89%) identified hurrying on stairs as a factor that would increase risk of falling; however, 63% stated that they do hurry on occasions, giving various reasons for this (multiple responses): answering a caller at the door (38%), answering the telephone (27%), retrieving items left upstairs (23%), and needing to use an upstairs toilet (13%). A sizeable proportion of participants (37%) said that they made a point of not hurrying for any reason. The older the participant the less likely they were to hurry (p < 0.001).

A large proportion of participants (92%) considered that carrying objects either up or down stairs could be hazardous, although only a small number (5%, n = 8), the majority of whom (n = 6) were age 75 and over, indicated explicitly that they no longer attempt to carry anything up or down stairs. Many participants mentioned strategies they employ to improve their safety when carrying objects up and down stairs, such as moving more slowly, resting items on steps and moving up or down a few steps at a time, and throwing items down to the bottom of the stairs. Some 29% of participants reported that when needing to move an object up or down stairs that may cause them a problem, they would not ask for help but "have a go anyway".

When asked about the frequency of use of handrails, 72% reported that they use handrails every time they use their stairs. In households with two handrails fitted, 74% of participants stated that they use both of these. Of those participants whose houses had one handrail, 24% said that they had considered having a second fitted. Of those who had not considered it, when asked, 17% thought that they would eventually need one.

In response to questions about use of stair lighting, 68% indicated that they avoid using stair lights during the day, with 18% stating that they do not switch the stair lights on if they need to get up and use the stairs during the night. Cost was mentioned by 11% of participants as a reason that they use lights to a minimum. This was in the context of 61% of households having illumination levels below the level of 50 lux during the day, where use of lighting would be recommended. Table 11.5 presents information on stair lighting and various other researcher observations regarding the stair environment.

In terms of items left on stairs, permanent objects were more likely to be positioned on stairs in detached houses, and temporary objects were seen more often on stairs in smaller terraced homes (p < 0.001). Examples of items seen on stairs are pictured in Figure 11.1 and Figure 11.2. When asked, 89% of participants identified leaving objects on stairs as a factor that would increase the risk of falling; however, a large proportion of

Table 11.5   Features of the Stair Environment

Households where artificial light required during the day (50 lux or less)	61%
Households with bulbs 60 Watt or less at top of stairs	44%
Low-energy bulbs in use at top of stairs	24%
Ambient illumination level at night, with lights switched off	Mean 0.04 lux
Households with objects placed on stairs	29%
Objects small and apparently temporary	10%
Objects apparently permanent (e.g., furniture)	19%
Households with worn, frayed, or poorly fitted stair coverings	29%
Households with loose rugs or mats positioned at top of stairs	11%

participants (71%) reported that they do place objects on stairs. Considerable overlap occurred between those identifying this as a risk factor and those engaging in the practice (p < 0.01). When asked whether their stair covering was in need of repair, 8% reported that it was. On inspection by the interviewer, however, 29% of the coverings were judged to be worn, frayed, or poorly fitted.

Figure 11.1   Temporary and permanent objects on stairs (household 095).

Figure 11.2 Object placed on half-landing with mirror on wall (household 009).

No differences were found with the social class of the participants, or with the tenure of the dwelling, in the proportions of households with poor stair coverings. Nor were there any differences between participants of different social class, or with different visual acuity, in the presence of loose mats or rugs at the top of stairs; however, it must be noted that the sample size for this analysis was small. It is also noteworthy that no one in the "oldest" age group had a rug or loose mat positioned at the top of stairs. Coverings in households where stairs are steep tended to exhibit more signs of wear, although this difference was not statistically significant. In some instances, heavily patterned carpets were found, which made it difficult to distinguish step edges (e.g., Figure 11.3a compared with Figure 11.3b).

In a small number of households (3%, n = 5), mirrors were found positioned on half-landings or close to the top of stairs, which might present a visual distraction (Figure 11.2). In other homes (6%, n = 9), stair lifts had been fitted along the length of the flight, thus reducing the width for other stair users.

Two-thirds of interviewees (66%) reported that they clean their stairs themselves, with 36% using a full-sized vacuum cleaner to do this. Proportionally more of the "oldest" age group chose to use a small hand-held vacuum cleaner, or do not clean the stairs themselves (p < 0.001). Of those who do clean their stairs, a majority (55%) reported doing so in

Figure 11.3   View looking down stairs. (Left) Household 030; (right) household 114.

the direction of top to bottom, moving down backward. Some 35% of the group who do their own cleaning identified various aspects of vacuum cleaner use as problems they thought would increase risk of falling.

Responses to questions concerning improvements made to the stair environment over the previous 5 years to make them safer found that 17% of households had installed a second handrail, whereas 16% had replaced the floor covering. Other changes reported by interviewees included: using brighter light bulbs, fitting more transparent lampshades, use of low energy light bulbs, and painting the walls a lighter colour, but in each of these cases, 5% or fewer of households had made the change.

11.4.3 Interviewee Health and Experience of Falls

Visual acuity was tested at a distance of 2.5 metres, with scores converted to conventional Snellen notation (Howarth and Bullimore 2005). Vision of 6/12 is considered adequate for most everyday tasks, whereas in the United Kingdom, the requirement for driving is approximately 6/15 or better. A high proportion of participants (68%) had uncorrected visual acuity worse than 6/12, with 19% of participants having corrected visual acuity worse than this. No significant differences were found between interviewees with and without history of falling on stairs with respect to their visual acuity. Visual stereopsis, one component of the depth percep-

Table 11.6   Use of Prescribed Medication

At least one prescribed medication taken daily	82%
Four or more prescribed medications taken daily	26%
Participants reporting that their medicines make them feel drowsy, dizzy, or affect their vision	16%

tion process was assessed using a Frisby stereotest (Frisby 1980). The results from this indicated that 50% were unable to detect the depth to a pattern with uncorrected vision, whereas 25% were unable to detect depth with vision corrected. No significant differences were found between participants with and without history of falling on stairs with respect to visual stereopsis.

Nearly all (99%) respondents reported using spectacles, with over half (57%) having bifocal lenses. A number of participants (16%) reported that their spectacles (all types) cause them visual problems when using stairs. These problems included an inability to judge depth (and consequent uncertainty in the location of steps) and distortion of the steps when using bifocal spectacles; however, many bifocal wearers (80%) reported that they did not experience any of these problems. Some 27% of spectacle wearers reported not wearing their glasses when needing to get up during the night.

As might be expected among this age group, use of prescribed medication was high (Table 11.6), with 16% of the sample reporting effects from their use of medicines that could affect risk of falling. A relationship was found between gender and the reporting of alcohol consumed in the previous 7 days before the interview, with 77% of men as opposed to 57% of women reporting having done so ($p < 0.05$). In addition, there was a difference with only 30% of men, compared with 53% of women, reporting avoiding alcohol while taking prescribed medication ($p < 0.05$). Although not prompted during the interview, a number of participants (10%) specifically mentioned using stairs as a form of daily exercise.

Details of interviewees' experiences of falling on either their own stairs or those in the homes of relatives are given in Table 11.7. No statistical relationship was found between the measures of functional ability (including age) and experience of falling on stairs. A positive association was found between experience of falling on stairs and taking four or more prescribed medications daily ($p < 0.05$) and problems holding onto the handrail (due to either design of rail or personal factors) ($p < 0.05$).

Approaching one-third of participants (28%) said they were concerned about falling on stairs, although 90% of participants rated their stairs as being safe, when asked about this separately. The majority of participants

Table 11.7   Falls on Stairs

Fallen at least once during previous 12 months	30%
Fallen three times or more during previous 12 months	8%
Led to attendance at hospital	20%
Presentation to general practitioner	13%
No medical attention sought	67%
Descending the stairs when fall occurred	66%
Carrying something at the time of the fall	26%
Hurrying at the time of the fall	10%
Footwear mentioned as an issue	24%

felt that their stairs were safe, regardless of their MMBI functional ability score or their ability to perform the "rise from stool" test.

At the conclusion of the interview, participants were asked if they had ever received any advice about safety on stairs. Among the respondents, 13% indicated that they had, most often from an occupational therapist or physiotherapist (7%), with 7% having found the advice given to them useful. No statistically significant associations were reported between those having received advice on stair safety and previous experience of falls. This was also the case for the indicators of physical capability (i.e., MMBI, grip strength, and rise from stool test).

11.5 Discussion

This study explored how older people use their stairs and how this is influenced by individual circumstances and behaviour. The results confirm that stair use does present a difficulty for a sizeable proportion of older people and indicate how the decisions and actions of this group affect their risk of falling in this location.

11.5.1 Interviewee Fall Experience

With 30% of the sample having fallen on their stairs during the past 12 months, the frequency is perhaps higher than might be expected, taking into account prevalence of falling in all locations (Lord et al., Chapter 3, this volume). This might indicate a sampling bias, with those having a reason to be interested in stair safety more likely to volunteer to participate in the research. If such a bias did exist, then it is possible that interviewees might be more apprehensive about using stairs than the wider population of older people, be in worse physical condition, or have less safe stair environments.

Among the variables examined for a relationship with falling, the relationship was non-significant for age, MMBI functional ability score, rise from stool test, visual acuity, and use of a walking aid. Significant associations between falling on stairs and use of four or more prescribed medications ($p < 0.05$) and problems holding onto the handrail ($p < 0.05$), are as might be expected. Other research has pointed to the possible involvement of footwear in falls on stairs (Startzell et al. 2000). In the present study, footwear was mentioned as an issue in 24% of the stair falls experienced by participants.

It is interesting that although those who had fallen previously on stairs rated themselves as being more likely to fall on stairs in the future ($p < 0.05$), a very high proportion of this group regarded their stairs as safe (87%). The implication of this is that those who have fallen on stairs have reduced confidence in their ability to use stairs safely for reasons other than the perceived safety of their particular stair environment.

11.5.2 Stair Environment

The majority of dwellings in this study (65%) had stairs steeper than the maximum allowed under current United Kingdom Building Regulations (42° maximum pitch). Some participants commented that they find it necessary to descend stairs with their feet positioned sideways, due to the size of the tread being insufficient to accommodate the full size of their feet. The Building Regulations recommend a minimum going of 220 mm for private housing, which Roys (2001 and Chapter 3, this volume) identified as smaller than the foot length of 95% of the adult population (or 100% if 30 mm is allowed for footwear). In the present study, the mean going of 214 mm was smaller than the minimum recommended. With average foot lengths of participants being 267 mm for men and 243 mm for women (unshod), it is clear that it is common for an individual's foot length to exceed the size of the steps on their stairs. Remedying this situation will be a long-term proposition, requiring attention to building codes and standards.

It is widely agreed that appropriately designed and fitted handrails are important (Templer 1992; Roys, Chapter 3, this volume). To prevent falls, the rail needs to be capable of taking the weight of a person when they pull on it. In a number of households visited during the survey (10%), the handrail was judged to be in need of repair. A third of homes (34%) had at some time had a second handrail fitted, a somewhat higher proportion than found by Edwards and Jones (1998). In their survey, less than 20% of individuals in a representative sample of persons age 65 and over had an extra stair handrail installed as an assistive device. Generally, interviewees with a second rail fitted to the stairs in the present study had lower functional ability ($p < 0.05$), suggesting that a need for additional support using stairs had been recognised. Interestingly, a high proportion of those with an additional handrail among Edwards and Jones' respondents (91%), and in our study (74%), reported using it, signifying that individuals find them beneficial. Interviewees often spoke favourably of the second rail, mentioning that it allowed them to use their stairs with increased confidence.

Although most participants reported their stair coverings to be in reasonable condition (86%), on inspection by the researcher, many (29%) were judged to be in need of replacement or repair. It appears likely that either the interviewees had a lower threshold for what constitutes reasonable condition, or else they had not noticed wear and tear that may have happened gradually over many years. Even if a householder is aware that a stair covering is in poor condition, there may be obstacles to having it replaced. These include the expense, difficulty involved in sourcing a replacement, disruption during fitting, or believing that awareness of any damage will be sufficient to avoid having an accident. Coverings in households where stairs are steep (i.e., the riser high and the going short) tended to exhibit more signs of wear. In these circumstances, the shoe heel may come into frequent contact with the carpet on the leading edge of the tread or the top of the riser, resulting in premature wear.

The pattern and colour of carpets may make it difficult to detect step edges in some circumstances (Figure 11.3a). Generally, coverings light in colour and non-patterned make steps more visible (Figure 11.3b), whereas those that are heavily patterned tend to camouflage the edges of steps (Howarth, Chapter 4, this volume). Part of this is due to the pattern making it difficult to judge depth and position, with patterns incorporating horizontal and vertical lines appearing to cause a particular problem (Cohn and Lasley 1985, 1990). These effects may be worse for those with poor visual acuity, poor visual stereopsis, or who do not wear their spectacles. Given the limited awareness of possible implications of colour and patterning, these issues are unlikely to be considered by older people when selecting stair carpets or other coverings for their homes.

Poor lighting has been suggested as a contributory factor in many fall accidents (Templer 1992), although precise lighting requirements are difficult to specify. Adequate lighting is necessary to see the steps on stairs and to detect objects. Moreover, dim lighting levels appear to be associated with poorer postural stability in older people (Brooke-Wavell et al. 2002). Light measurements taken during the day in participants' homes, with artificial lighting switched off, found 61% of households to have light levels of 50 lux or less. Recommendations for the workplace specify this level of illumination as appropriate for areas visited infrequently, requiring limited perception of detail (Howarth 2005). In the present survey, many of the homes visited were fitted with light bulbs that were less than 100 W (44%). Coupled with this, some households had lampshades of a design that further restricted the levels of lighting. A number of participants had low-energy, low-wattage light bulbs at the top of stairs. This type of luminaire requires changing much less frequently than a traditional incandescent lamp, thereby reducing the possibility of falling when changing the bulb. A second advantage is reduced cost of operation, allowing people to leave stair lights on without the worry of large electricity bills. Cost was mentioned by 11% of participants as a reason that they keep lighting to a minimum.

Although smoke alarms were not examined explicitly in the environmental survey, a number were seen positioned over the stairs. In these circumstances, it may be difficult for the householder to test the alarm or change the battery, without putting him or herself at risk of falling. Guidance and standards for installation may be needed to address this problem.

Items placed on stairs can be a trip or slip hazard, or may form an injurious object to fall against. Objects may be positioned on stairs on a temporary or more permanent basis, with many instances of both found in the homes of participants in this study. The finding that items such as furniture were more likely to be positioned on stairs in larger, detached dwellings, and temporary objects were seen more often on stairs in terraced houses (p < 0.001), may reflect the availability of space. Stairways in larger houses are more likely to have half-landings of sufficient size to accommodate larger, permanent objects. Storage space may be restricted in smaller homes, necessitating more frequent movement of items between levels or use of stairs as an overspill storage area. The potential exists for housing design to more adequately anticipate home occupiers' needs for storage and inclinations for furnishing, to address this problem.

11.5.3 Stair Usage

Half of all participants said that there is no particular time of day that they use their stairs more often than any other. Over one-third (38%) reported the morning to be the period of most frequent use, giving various reasons for this. Some individuals mentioned the effects of medication increasing the need to use the toilet in the mornings, this sometimes being urgent, making it necessary to hurry (one-third of the households had only a single toilet upstairs). Other reasons given for using stairs more in the morning were to perform housework tasks or to get ready to go out.

As might be expected, individual functional ability affects stair use, with those with decreased ability using stairs less often. Of those who stated that they use stairs infrequently, respondents described conditions that affect their legs, (e.g., pain in knees or hips), or other medical problems that make it difficult for them to exert themselves (e.g., heart problems or breathlessness), as the reason for their stair avoidance. Perhaps surprisingly, this study found no indication of cohabitation affecting patterns of stair use, as has been suggested elsewhere (Smith et al. 1994). It might be expected that in some circumstances, a more mobile partner will use the stairs more frequently to assist a less able spouse.

Similar to Startzell et al. (2000), this study suggested that older people seem to use stairs with increased caution in some respects, while still engaging in other potentially dangerous behaviours, such as leaving objects on stairs. The discrepancy manifests itself with the majority of interviewees identifying hurrying on stairs, carrying items and leaving objects on stairs as likely to increase risk of falling, but with many continuing to engage in these activities. Almost a third of participants in the study (29%), for example, said that they would still attempt to carry an object up or down stairs that might cause them difficulty. The findings from our focus group research (Haslam et al. 2001) suggested that although individuals recognise certain activities as hazardous, they continue to practise them through perceived necessity and a view that the manner in which they personally perform the function reduces any risk to an acceptable level. Further possible explanations include the desire to maintain independence or a lack of access to someone who is able to provide assistance. Given the evidence from other research (Templer 1992) that this conduct does increase risk of falling on stairs, there would appear to be a need to raise awareness of the risk and of the potential consequences.

With regard to use of lighting (as opposed to provision of lighting as discussed in Section 11.5.2), there was a conflict between recognition of the importance of good stair lighting and the low usage of supplementary lighting in practice (68% not using stair lights during the day and 30% generally using lighting to a minimum). Using stairs at night without switching on lighting appears undesirable from a safety perspective (Connell and Wolf 1997), a point that most interviewees recognised. Almost one-fifth of interviewees (18%), however, reported not switching on lighting if they needed to use their stairs after retiring to bed. A variety of explanations were given for this, including not wanting to disturb a sleeping partner, presence of sufficient illumination from outside, eyes already adapted to the dark, and individuals being familiar with their own stairs. When night illumination readings were taken in a selection of households, levels were unmeasurable with the lights switched off. Howarth (Chapter 4, this volume) describes how visual performance is reduced in such conditions. Although the need for education appears obvious in this respect, there might also be design solutions that would improve safety. For example, there may be benefit from the provision of lighting around stair areas, activated by movement, gradually increasing to an optimum level. Clearly, the technology would need to be of adequate sophistication to be acceptable to home occupiers and research involving users would be needed to achieve this.

Cleaning on and around stairs appears to present particular problems, due to a combination of difficult access (e.g., landing windows) or the need to use awkward and heavy equipment (vacuum cleaners in particular). Some older people seem to reach a point where reduced strength and flexibility lead them to seek alternative methods, such as using smaller, hand-held, battery operated vacuum cleaners, or employing a relative or another person to clean on their behalf. It appears plausible, however, that some individuals will pass through a stage before this, where they struggle to clean as in the past, not accounting for their changing abilities, placing themselves at increased risk as a consequence. The challenge then is to encourage older people to appreciate their limitations, without instilling negative attitudes. Again, the possibility exists for improved design to facilitate cleaning. In one extreme example found in the household survey, a window was installed over an open stairwell (Figure 11.4). The participant gave the following description of how she gained access to clean and change the curtains: "I have to place planks of wood from the upper banister to the window sill and then walk across to the window. I don't feel at all secure when I do this." Connell and Wolf (1997) reported similar examples of older individuals engaging in hazardous activities, which would require a very high level of physical ability to do so safely.

11.5.4 *Individual Capability*

The age profile of the sample of older people participating in the survey was similar to the distribution for those age 65 and over in the United Kingdom. Apart from the possible implications of the sample recruitment (Section 11.5.1), there is no evidence of the sample being atypical of the general population in any other systematic respect.

Use of prescribed medication and alcohol, either alone or in combination can affect reaction time, balance, and judgement. This applies particularly to psychotropic medication (i.e., as used to treat conditions such as anxiety, depression, and insomnia) (AGS et al. 2001). The high rate of use of prescribed medication in the survey, with 82% of those interviewed taking at least one medication daily and 26% taking four or more, is not unusual among those age 65 and over (Cumming et al. 1991). Neither are the unwanted side effects, such as drowsiness, dizziness, or

Figure 11.4   Landing window located above stairs (household 132).

effects on vision, reported by 16% of respondents. Indeed, improved prescribing has been highlighted as an important measure for fall reduction among older people (AGS et al. 2001; NICE 2004).

Again, the widespread use of alcohol reflects general societal behaviour. Of course, it requires a certain level of alcohol imbibition to affect risk of falling, although few studies have sought to quantify this, nor considered any differences that might arise with ageing. Over one-third (38%) of participants reported drinking alcohol when taking prescribed medication, giving reasons such as not drinking enough to affect their medication or waiting for a time after taking medication before consuming any alcohol. A significant proportion of the sample, at least one-third, said that they had not received any advice from their doctor or pharmacist in this respect.

It appears there may be scope to improve individuals' knowledge of when they personally could be at increased risk of falling due to medication or through mixing medications with alcohol. This might be achieved through improved communication of medicine information and instructions to older people and their caregivers regarding use and possible side effects, and by raising awareness of the extent to which alcohol consumption may contribute to falls among this age group (Wright and Whyley 1994).

Nearly all (99%) the respondents reported using spectacles, with assessment of vision confirming the need for this. Even when wearing spectacles, 19% of participants had corrected visual acuity less than 6/12, a level considered adequate for everyday tasks. The loss of stereopsis, indicated by the 25% of the sample unable to detect depth to a pattern with corrected vision (50% with vision uncorrected), may impair the negotiation of stairs (Haegerstrom-Portnoy et al. 1999). Although it is not possible to draw definitive conclusions that deteriorating vision is a contributory factor in falls on stairs, it appears likely that this will be the case (Startzell et al. 2000), especially when negotiating unlit stairs at night without using spectacles (27% of participants reported not wearing their spectacles when moving around their home during the night).

A research question raised by this study has been "To what extent could bifocal and varifocal spectacles be a contributory factor in falls on stairs?" Of participants who wore bifocal spectacles, 20% reported that these caused problems when descending stairs. Problems included distortion and difficulty judging depth, with consequent uncertainty in locating steps. A subsequent study by Davies et al. (2001), examining records in a preexisting injury database, found a significant association between falls on stairs, where the first event was the missed edge of a step and the wearing of bifocal or multifocal lenses.

Various methods were reported to be used to reduce problems due to wearing bifocals when descending stairs, such as positioning the head so that the wearer is not looking through the portion of the lens causing visual distortion; not looking at the steps, but looking ahead to the bottom of the stairs; and not wearing spectacles at all when using stairs.

Some participants (27%) reported that their opticians or optometrists had warned them of the need for care when using stairs and, in any case, for stronger prescriptions, the distorting effects on vision are readily apparent. Participants mentioned that they adapted to wearing bifocals, reporting that single steps and street curbs cause them the most problems subsequently; however, whether this compensatory behaviour actually occurs and, if so, whether it is sufficient to prevent falls is unknown.

The role of exercise in improving strength and fitness among older people has been established by several studies (Lord et al., Chapter 5, this volume), and individually tailored exercise programmes have been shown to lead to a reduction in falls (Gillespie et al. 2003). Although information on exercise was not collected explicitly from interviewees during the present study, a small proportion of participants (10%) indicated their view that using stairs could be beneficial in this respect. Interviewees also gave accounts of using stairs for this purpose, when perhaps their health might make this appear misguided. The pros and cons in this respect will vary from one individual to another.

Although moving to a bungalow will inevitably reduce the risk of falling on stairs, it emerged from the focus group study that stairs and steps in shops or friends' homes may then become more difficult to negotiate. The term "bungalow legs" was used by one person to describe the difficulty and aches and pains experienced on stairs after a period of living in a holiday bungalow (Haslam et al. 2001). There may also be a psychological effect accompanying reduced stair use, leading to increased apprehension on the occasions on which they are used.

11.5.5 Awareness of Risks

Although the study participants were often able to appreciate and understand situations that might increase risk of falling on stairs, they generally needed to have these brought to their attention first. This suggests that the risks are fairly evident, but not so evident that they will necessarily be noticed before a problem (fall or near miss) has occurred. This suggests a need for raising awareness. Unfortunately, only around 1 in 10 of those interviewed were able to recall ever having been given any advice on stair safety (13%), with half of these individuals receiving guidance from an occupational therapist or physiotherapist.

Our anecdotal experience indicates that older people are willing to receive and accept advice, interested to learn how to minimise the likelihood of injuring themselves. This is supported by a strong desire to maintain independence and autonomy. Routes for education are many and approaches need to be tailored to different circumstances. It may be appropriate to begin the process at the time people retire, when most are still active enough to be able to make changes to their physical environment for themselves (Smith et al. 1994). As might be expected, personal experience, involving either the individual or a close associate, appears to have the strongest effect on perception of safety. It is also suggested that discussion or drama-based forums might prove a useful component of educational activities.

11.6 Concluding Discussion

This investigation used focus groups and a home interview survey to examine patterns of stair use among older people and the presence of factors likely to affect risk of falling. Of particular interest was the influence of behaviour on exposure to fall risk. Similar to Connell and Wolf (1997), the research demonstrates pathways through which the actions and decisions of older people affect their environment, themselves, and the interaction between the two. Drawing together the evidence from the focus group and interview research, three routes are proposed through which the behaviour of older people influences their safety on stairs (Figure 11.5):

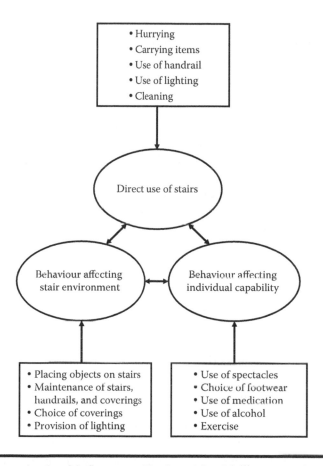

Figure 11.5 Behavioral influences affecting risk of falling on stairs.

1. Behaviour involved in direct use of stairs
2. Decisions and actions affecting the stair environment
3. Behaviour affecting individual capability

It is argued that efforts aimed at reducing falls on stairs among older people should give attention to each of these three areas.

Relevant to this assertion, a review led by the American Geriatrics Society (AGS), which examined the efficacy of interventions aimed at preventing falls among older people, concluded that multifactorial interventions including a behavioural change component demonstrated benefit (AGS et al. 2001). No evidence, however, supports the use of behaviour modification approaches when used in isolation (AGS et al. 2001; Gillespie et al. 2003), although this judgement is based on only a small number of studies. It should be recognised that attempts to change long-established habits and routines, especially where there are a range of these habits and routines to be targeted, presents a significant challenge (Connell and Wolf 1997), and depends heavily on the mode of delivery. There appear to have been no studies that have examined explicitly the effectiveness of behavioural approaches to improving safety with regard to falls on stairs. Further research is desirable in both of these respects. The evidence on the value of attending to environmental hazards appears clearer, with several intervention evaluations including a substantial hazard modification component, reporting a reduction in falls (Gillespie et al. 2003).

Notwithstanding the findings from the various research studies, the recommendation from the National Institute for Clinical Excellence (NICE 2004) that a participative approach should be taken, encouraging the involvement of older people in fall prevention programmes, is appealing. There is an empowering role for older people and their caregivers in assuming responsibility for making their environments safer and developing their individual capabilities. Other groups also have important contributions to make with regard to safety on stairs. Medical and social advisors have a role in raising awareness of factors affecting stair safety and providing advice on what can be done to alleviate risks. Looking to the future, home designers have a responsibility for providing stairs that meet the requirements of older users and their differing needs and capabilities.

Although single strand approaches to prevention are necessary to collect evidence on the effectiveness of different fall prevention strategies, the appropriateness of this for wider practice is questionable. The other chapters in this book have demonstrated the manner in which falls arise from a complex interaction of circumstances. Falls are a multifactorial problem, requiring a multifactorial approach to prevention, especially among older people, where personal factors are so heavily involved.

Acknowledgments

The U.K. Department of Trade and Industry (DTI) funded this research. Any opinions and conclusions expressed, however, are those of the authors alone and do not necessarily reflect DTI policy.

References

Allied Dunbar National Fitness Survey, 1992. *A report on activity patterns and fitness levels: main findings* (Sports Council and Health Education Authority: London).

American Geriatrics Society (AGS), British Geriatrics Society, and American Academy of Orthopaedic Surgeons Panel on Falls Prevention, 2001. Guideline for the prevention of falls in older persons. *Journal of the American Geriatrics Society*, 49, 664–672.

Askham, J., Glucksman, E., Owens, P., Swift, C., Tinker, A., and Yu, G., 1990. *A review of research on falls among elderly people* (Department of Trade and Industry: London).

Bath, P.A. and Morgan, K., 1999. Differential risk factor profiles for indoor and outdoor falls in older people living at home in Nottingham, U.K. *European Journal of Epidemiology*, 15, 65–73.

Brooke-Wavell, K., Perrett, L.K., Howarth, P.A., and Haslam, R.A., 2002. Influence of the visual environment on the postural stability in healthy older women. *Gerontology*, 48, 293–297.

Cohn, T.E. and Lasley, D.J., 1985. Visual depth illusion and falls in the elderly. *Clinics in Geriatric Medicine*, 1, 601–620.

Cohn, T.E. and Lasley, D.J., 1990. Wallpaper illusion: cause of disorientation and falls on escalators. *Perception*, 19, 573–580.

Connell, B.R. and Wolf, S.L., 1997. Environmental and behavioral circumstances associated with falls at home among healthy elderly individuals. *Archives of Physical Medicine and Rehabilitation*, 78, 179–186.

Cumming, R.G., Miller, J.P., Kelsey, J.L., Davis, P., Arfken, C.L., Birge, S.J., and Peck, W.A., 1991. Medications and multiple falls in elderly people: the St. Louis OASIS study. *Age and Ageing*, 20, 455–461.

Davies, J.C., Kemp, G.J., Stevens, G., Frostick, S.P., and Manning, D.P., 2001. Bifocal/varifocal spectacles, lighting and missed-step accidents. *Safety Science*, 38, 211–226.

Delbaere, K., Crombez, G., Vanderstraeten, G., Willems, T., and Cambier, D., 2004. Fear-related avoidance of activities, falls and physical frailty: a prospective community-based cohort study. *Age and Ageing*, 33, 368–373.

Department of Trade and Industry (DTI), 2003. *24th (final) report of the home and leisure accident surveillance system* (Department of Trade and Industry: London), DTI ref 03/32.

Dowswell, T., Towner, E., Cryer, C., Jarvis, S., Edwards, P., and Lowe, P., 1999. *Accidental falls: fatalities and injuries — an examination of the data sources and review of the literature on preventative strategies* (Department of Trade and Industry: London), DTI ref. 99/805.

Edwards, N.I. and Jones, D.E., 1998. Ownership and use of assistive devices amongst older people in the community. *Age and Aging*, 27, 463–468.

Experian, 1999. Personal communication: Socioeconomic class derived from postcode analysis. Experian, Nottingham.

Frisby, J.P., 1980. The Frisby stereotest: amended instructions. *British Orthoptic Journal*, 37, 108.

Gillespie, L.D., Gillespie, W.J., Robertson, M.C., Lamb, S.E., Cumming, R.G., and Rowe, B.H., 2003. Interventions for preventing falls in elderly people (Cochrane Review). In: *The Cochrane Library*, Issue 4, 2003 (Wiley: Chichester).

Haegerstrom-Portnoy, G., Schneck, M.E., and Brabyn, J.A., 1999. Seeing into old age: vision function beyond acuity. *Optometry and Vision Science*, 76, 141–158.

Haslam, R.A., Sloane, J., Hill, L.D., Brooke-Wavell, K., and Howarth, P., 2001. What do older people know about safety on stairs? *Ageing and Society*, 21, 759–776.

Healy, J. and Yarrow, S., 1998. *Safe at home? Views of professionals on preventing accidents in the home among older people* (Health Education Authority: London).

Howarth, P.A., 2005. Assessment of the visual environment. In: *Evaluation of human work* (edited by Wilson, J.R.), 3rd ed., pp. 663–692.

Howarth, P.A. and Bullimore, M.A., 2005. Vision and visual work. In: *Evaluation of human work* (edited by Wilson, J.R.), 3rd ed., pp. 573–604.

Lord, S.R., Sherrington, C., and Menz, H.B., 2001. *Falls in older people: risk factors and strategies for prevention* (Cambridge University Press: Cambridge).

Masud, T. and Morris, R.O., 2001. Epidemiology of falls. *Age and Ageing*, 30-S4, 3–7.

Nagata, H., 1993. Fatal and non-fatal falls — a review of earlier articles and their developments. *Safety Science*, 16, 379–390.

National Institute for Clinical Excellence (NICE), 2004. *Falls: the assessment and prevention of falls in older people* (National Institute for Clinical Evidence: London), Clinical Guideline 21.

Office for National Statistics (ONS), 1999a. Estimates for population figures, 1991 census data. Office for National Statistics, London.

Office for National Statistics (ONS), 1999b. Estimated housing stock, 1997–1998. Survey of English Housing, Office for National Statistics, London.

Pauls, J.L., 1985. Review of stair-safety research with an emphasis on Canadian studies. *Ergonomics*, 28, 999–1010.

Pauls, J.L., 1991. Safety standards, requirements, and litigation in relation to building use and safety, especially safety from falls involving stairs. *Safety Science*, 14, 125–154.

Prudham, D. and Evans, J.G., 1981. Factors associated with falls in the elderly: a community study. *Age and Ageing*, 10, 141–146.

Roys, M., 2001. Serious stair injuries can be prevented by improved stair design. *Applied Ergonomics*, 32, 135–139.

Smith, D.B.D., 1990. Human factors and ageing: an overview of research needs and application opportunities. *Human Factors*, 32, 509–526.

Smith, D.W.E., Snell, J., Brett, A.W., Jackson, F.W., Straker, J.K., and Ulmer, M.E., 1994. A study of stairs in the housing of independently living elderly people. *International Journal of Aging and Human Development*, 39, 247–256.

Startzell, J.K., Owens, D.A., Mulfinger, L.M., and Cavanagh, P.R., 2000. Stair negotiation in older people: a review. *Journal of the American Geriatrics Society*, 48, 567–580.

Templer, J., 1992. *The staircase* (MIT Press: Cambridge, Massachusetts).

Tinetti, M.E., Speechley, M., and Ginter, S.F., 1988. Risk factors for falls among elderly persons living in the community. *New England Journal of Medicine*, 319 (26), 1701–1707.

Wright, F. and Whyley, C., 1994. *Accident prevention and risk-taking by elderly people: the need for advice* (Age Concern Institute of Gerontology: London).

Chapter 12

Preventing Falls

Roger Haslam and David Stubbs

CONTENTS

12.1 Overview

The early chapters in this book described the mechanisms of human gait, including discussion of falls in different circumstances. It was explained that walking is a complex feat of biological engineering, involving postures that are inherently unstable. Separate attention was given to our visual sense, in view of the significance of this for both monitoring the environment and maintaining balance. Processes involved in slipping and tripping have been described, together with an account of the interaction between footwear and flooring and the manner in which this is affected by the presence of substances such as liquids and ice.

The two chapters (Chapter 3 and Chapter 5) examining steps and stairs and falls among older people considered situations where falls are a particular problem. Steps and stairs are common features of the built environment and are a concern because falls in these locations lead to injuries of greater severity than those from falls on the level. Older people are particularly prone to falling due to changing abilities with ageing and they take longer to recover than younger adults and children.

Approaches to fall investigation have been reviewed from both a research and practical perspective. A range of methods, from epidemiological analysis of injury databases to investigation of individual incidents, are of value in understanding factors involved in falls among different groups and in different settings. Separately, the different methods described each have their limitations, but when used in combination, they allow a detailed picture of causation to be established. Fall investigation as part of risk management practices needs to consider prevailing attitudes held in the wider community toward falls. Often, these reflect a view that falls are either inevitable or in large part due to the recklessness of those involved. Structured investigation of individual falls contributes to organisational learning as well as the identification of particular hazards in an incident that ought to be dealt with.

The series of case studies presented in Chapters 8–11 elaborated on the variety of circumstances in which falls occur and approaches to their prevention, drawing upon practical examples. These encompassed factors involved in fall causation of both environmental and behavioural origin.

12.2 The Politics of Falls

Readers may have discerned from the contributions to this book a sense of the "politics" surrounding falls and the effects of this on the direction research and preventative activities have taken. Although this book has emphasised the need for a multidisciplinary approach to understanding and preventing falls, including consideration of the wider circumstances affecting exposure to risk of falling, mention was made in Chapter 1 of the distinct and largely independent attention given to falls by practitioners, researchers, and policymakers from different backgrounds. Falls in the workplace and falls among older people are contrasting examples in this respect.

In the workplace, greatest effort appears to have been devoted to developing methods for the measurement of slip resistance in different situations and to achieving technological improvements with flooring and footwear. Less attention has been given to the many other factors involved in workplace falls, especially those intrinsic to the person, such as health, fitness, strength, balance, and coordination, as well as influences upon these. Moreover, surprisingly few intervention trials of any rigour, aimed at establishing what works and what does not work for preventing falls in the workplace, have been conducted. At present, a notable absence of objective evidence is available in this respect. This is in contrast to fall prevention among older people, where a significant number of intervention trials have now been undertaken, with the findings from these individual trials then collated and subjected to systematic review.

Notwithstanding the efforts directed at evaluating intervention approaches, the focus of fall prevention among older people has perhaps been narrower than desirable. Efforts to address the serious problem of falls among this group have, understandably enough, been led by the medical community, bringing a medical perspective to the issue. Indeed, the increasing emphasis on evidence-based, preventative medicine over recent years is at least partly responsible for the concerted attention that the problem of falls among older people is now receiving. Medical practitioners and researchers perhaps have a tendency to view falling as an illness, however, with an aetiology predominantly located with the person. An emphasis on medical assessment and treatment, with a primary focus on healing and avoiding reoccurrence, may skew attention away from efforts aimed at preventing initial falls happening in the first place.

It is argued here that in the case of both falls in the workplace and falls among older people, greater recognition is desirable of the wide-ranging precursors to falls, with similarly wide-ranging approaches to prevention. It is also surprising, especially to those familiar with user-centred, participative approaches to addressing problems involving people, the extent to which individuals appear to be viewed as passive recipients of fall prevention initiatives. Efforts aimed at reducing falls must recognise that their success will be influenced materially by the knowledge, attitudes, and beliefs of those toward whom fall prevention measures are directed, thus determining acceptance and resulting behaviour.

Another manifestation of the politics of falls may be observed in the high profile of flooring and footwear manufacturers that are eager to secure commercial benefit for their products. This applies similarly to manufacturers and retailers of floor cleaning agents, cleaning equipment, and treatments to increase slip resistance. When it comes to methods and devices for measuring the frictional properties of flooring, discussion has been equally intense, with large numbers of alternative techniques proposed over the last 50 years. As yet, improvements still need to be made in providing a slipperiness measurement method that better accounts for the dynamic processes involved in the interaction between foot and floor, while being suited to practical use. A considerable proprietary advantage can be gained for developers of a product or method adopted for widespread use. One outlet for these various interests is through participation and lobbying of the committees working on building standards and regulations. Current specifications for the dimensions of steps and stairs, for example, are in part a compromise between those arguing for the needs of the user and builders and clients concerned with construction costs.

On another front, an increased awareness of falls as a serious problem has been accompanied by increased levels of litigation from fall victims seeking compensation for their injuries. The defendants in such cases, employers, and those responsible for facilities used by the public are increasingly exposed if they are unable to demonstrate that they have taken appropriate measures to prevent falls occurring. Individual claims present interesting questions on the detailed dynamics of falling, the relationship between these and injury outcomes, and the extent to which an individual's actions might have been contributory. In each of these respects, more can be learned about falls, but notwithstanding this, substantial compensation awards have been made to fall victims who have had the misfortune to experience serious injury. A positive outcome of such litigation has been the extent to which it has acted as a forceful stimulus for more widespread implementation of safety measures aimed at fall prevention.

12.3 Understanding Falls

It has been discussed throughout this book that falls arise from a combination of environmental and personal factors that, in turn, are affected by individual behaviour and organisational influences (Figure 12.1). Hazards present in the environment, such as slippery flooring or an uneven floor surface, interact with the ability of individuals to detect such hazards, take avoiding action, and to recover balance should this be disturbed. Detection, avoidance, and balance recovery are affected by individual health, fitness, and alertness, with these influenced by the extent to which people exercise or suffer from illness and fatigue. Psychotropic medication, in widespread use for the treatment of conditions such as anxiety, depression, and insomnia, can affect coordination and concentration. In these respects, alcohol has similar effects that are readily apparent.

The extent to which fall hazards in the environment give rise to problems is affected by individual behaviour. An individual, in a hurry, carrying a heavy or bulky object, or giving their attention to something else (e.g., using a mobile phone), may be less likely to notice a slippery surface or a low step. Awkward or bulky loads may impede vision and make it more difficult to recover balance in the event of a slip or trip. Other behavioural influences arise from the choices individuals make in terms of using appropriate footwear, use of lighting, and, where necessary, use of correct spectacles. In some circumstances, individuals might act in

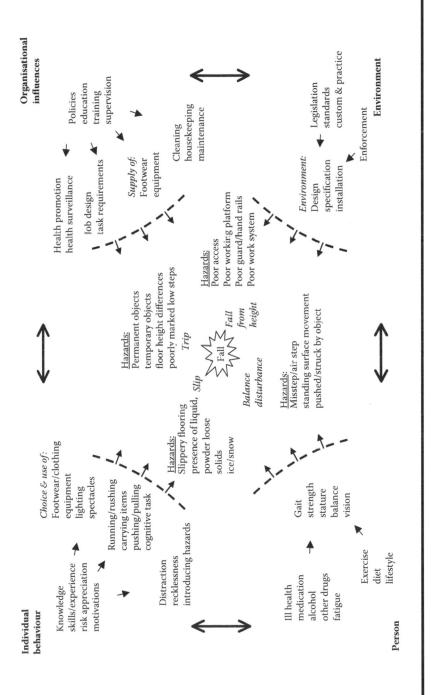

Figure 12.1 Factors and influences leading to falls.

a reckless fashion, venturing into unsafe situations. Examples commonly involved in falls from height are the use of steps and ladders in inappropriate circumstances and workers venturing onto fragile roofs. Such behaviour is affected by the knowledge of individuals, their appreciation of risk, and other factors that motivate them.

Without the presence of hazards in the environment, the majority of falls would not occur. The existence of hazards depends on the design of buildings and walking routes; specification of flooring surfaces, steps, stairs and other means of access to height; provision of lighting; and then the adequacy with which each of these is installed. These factors are affected by custom and practice, legislation, standards, and building codes, as well as the extent to which the latter are regulated and enforced. Organisational influences from those providing services to the public and employers bear upon individuals, the activities they perform, and the conditions in which these are undertaken. Practices at community level with regard to health promotion and medical care have an effect on the health and fitness of individuals, especially older people. Those responsible for the built environment have an obligation to ensure that areas are cleaned and maintained in satisfactory condition. Employers have a duty to provide footwear and work equipment appropriate to a particular workplace. Organisational policies affect each of these, as does the education, training, and supervision that are in place.

The consequences of a fall can include both physical injury and psychological distress, with wide-ranging outcomes. The severity of injuries depends upon, among other things, the vertical distance through which a fall occurs and the hardness of the surface on which the person lands. The outcome may range from bruising and grazes, of little long-term effect, through to fractures requiring a long period to heal, leaving a person with substantially reduced mobility and function in the interim. Falls can also lead to persistent back problems or, in the worst instances of spinal injury, quadriplegia. A blow to the head from falling, if not fatal, may lead to serious brain injury. In the worst cases, an unexpected fall can be a life-changing event, with severe repercussions for both the individual and those around them. Among older people, even a noninjurious fall can leave individuals with a loss of confidence to the extent that restricted activity ensues. Thus, the results of a fall can vary from being of little consequence through to being extremely serious.

12.4 Preventing Falls

The multifaceted problem of falls requires a multifactorial approach to prevention. This can be structured as having three generic components (Table 12.1):

1. Primary prevention
2. Residual risk reduction
3. Measures to maximise individual capability

We have not sought to separate discussion of fall prevention for different groups and differing circumstances, on the basis that conventional boundaries adopted between different categories of fall and fallers conceal a considerable overlap of issues. Older workers, for example, face both the hazards present in the workplace, along with gradual onset of the changes accompanying ageing. Falls from height may arise from similar antecedents to those that cause falls on the level and on steps and stairs.

12.4.1 Primary Prevention

Primary prevention of falls aims to eliminate fall hazards at source, through the design of the built environment and work/activity systems. Flooring should offer appropriate slip resistance for the different conditions to which it will be subjected. Similarly, walkways and walking areas should be designed and constructed to avoid trip hazards. Steps and stairs should be conspicuous and fitted with handrails of a design allowing a satisfactory grasp. In addition to consideration of the specification of flooring and stairways, primary prevention involves attention to the equipment used (e.g., to avoid spillages and other walkway contamination), the manner in which equipment is arranged, the tasks people need to perform, and the extent to which each of these might affect the risk of falling. Provision of sufficient storage is a measure aimed at reducing the need for objects and materials to be placed in the walking path, which may then present a trip hazard.

Permanent means of access to height should be provided to avoid the need for use of portable steps and ladders. This should be the case even when access will only be infrequent, as might occur with areas used for longer-term storage or where there might be a need for maintenance, for example. It is desirable, where possible, to avoid the need for people to stand or walk on surfaces that might move unpredictably, as is sometimes the case with public transport. Where this is unavoidable, grab-rails and other handholds should be provided. Handholds are also beneficial where people are required to perform awkward movements, getting into or out of a bathtub, or for entry/egress from a vehicle, for example. The presence of adequate lighting is important to allow people to monitor the walking surface and detect irregularities and other problems. Visual distraction should be avoided in situations where it is important that individuals view the walking surface, as is the case when approaching steps, stairs, or escalators.

Table 12.1 Fall Prevention Measures

Primary Prevention	Risk Reduction	Maximise Capability
Provide slip-resistant flooring	Provide education and awareness raising of fall risks and fall consequences	Promote and monitor health among vulnerable groups
Design work/activity systems to avoid presence of fall risks (attention to environments, equipment, layouts, tasks, and people)	Perform fall risk assessments and implement controls	Encourage exercise for strength, coordination, and balance
Cover outside walkways to keep off rain, snow, ice, and leaves	Organise sustainable housekeeping procedures for inspection, cleaning, and maintenance	Encourage and facilitate good diet
Design walkways to exclude trip hazards	Manage fall risks introduced during installation, cleaning, and maintenance	Adopt medication prescribing protocols to minimise fall risk
Provide sufficient, convenient space for storage	Provide warning signs for slip hazards	Promote behavioural sleep management programmes to address insomnia
Avoid presence of low steps	Mark trip hazards	Discourage use of alcohol where significant fall risks are present
Install steps and stairs of appropriate dimensions	Provide durable, remedial marking of step edges	Encourage use of suitable footwear
Provide step edges with good contrast	Fit additional handrails	Encourage use of suitable clothing
Install handrails	Fit additional grab rails	Encourage eye tests and appropriate use of spectacles
Avoid visual distraction in step/stair locations	Fit barriers for edge protection	
Provide permanent access to areas at height	Encourage use of lighting	
Avoid need for walking/standing on surfaces that move unpredictably	Discourage carrying of awkward, heavy loads	
Install grab rails and hand holds	Avoid creating circumstances that encourage rushing	
Install adequate lighting	Implement risk management protocols for inclement weather	
Design and select environment features to facilitate cleaning and maintenance	Implement risk management protocols for those at increased risk of falling	
Design and select environment features for durability and resistance to damage	Provide assistive mobility aids for those in need	

It should be remembered that walking surfaces and pathways will need to be cleaned and maintained, and the design and installation should make allowance for this. In addition, to avoid hazards being introduced by wear and tear, installations should be appropriately durable and resistant to damage. Pedestrian walkways can be protected from vehicle damage, for example, by ensuring a physical separation between the two (e.g., through installation of bollards).

12.4.2 *Risk Reduction*

Even with concerted attention to primary fall prevention, it is inevitable that fall hazards will continue to be present in the environment. Risk reduction measures are, therefore, necessary to reduce the likelihood of falls arising from these hazards. An important starting point is to raise awareness of the problem and, through education, promote understanding of risk factors for falling and how they can be ameliorated. Accompanying this is a need for risk assessment and risk management.

Where it is possible that slip or trip hazards might arise in an area used by pedestrians, it is important that adequate procedures are implemented to detect these and to remedy the situation. Indoor flooring will usually need to be cleaned periodically, partly for the sake of appearances, but also to remove dirt and debris that may present a risk of falling. During the cleaning process, it is possible that other fall hazards might be introduced (e.g., the risk of slipping with wet vinyl or tiled floor surfaces while these are drying). With regard to maintenance, routine inspection programmes should be arranged for walking areas and pathways. In all cases, housekeeping procedures should be designed to be sustainable, so that initial good practices do not deteriorate to the point of becoming ineffective, as is sometimes the case.

Where fall hazards are present and cannot be removed immediately, an obvious action is to warn of their existence. This can be through use of signage warning of a risk of slipping, or by marking step edges, for example. In circumstances where stairs have to be negotiated or where movements of a difficult nature need to be performed, then it may assist some, if not all users, to fit additional hand- or grab-rails. An important measure to prevent falls from height is to install barriers on edges and around areas open to a lower level. These guardrails may be permanent or temporary, depending on the situation. A primary prevention measure is to install sufficient lighting; however, this will only be effective if the lighting is actually used. Both carrying items and hurrying are other behavioural factors contributing to falls and should be discouraged in circumstances where other fall risk factors are present.

The risk of falling increases in certain situations. Poor weather, resulting in outdoor areas becoming covered with ice or snow, is frequently accompanied by increased prevalence of falls, unless appropriate precautions have been taken. It should be possible to plan ahead for such occasions and authorities and employers ought to be ready and prepared to implement measures to reduce risk, either through clearing affected areas or by reducing exposure to the slippery conditions (e.g., by encouraging people to stay indoors). Certain individuals may also be at an increased risk of falling, either due to frailty or a medical condition. Again, risk management protocols should be in place to reduce, as far as possible, the risk of injury to such persons. Mobility aids and personal protection, such as hip protectors, should be provided if assessment indicates that these will be beneficial.

12.4.3 Maximise Capability

A third strand of the fall prevention process is to endeavour to maximise individual ability to cope with the challenges present in negotiating our everyday environment. Clearly, this applies more to some groups of the population than others. An overarching activity should be to seek to promote and monitor health among vulnerable groups. Encouraging exercise to increase and maintain strength and coordination can help to improve balance. Diet can be beneficial in encouraging bone strength as well as contributing to general health.

As already mentioned, certain medications can affect balance and concentration, and good evidence indicates an association between polypharmacy and risk of falling. The onus is on prescribers to consider fall risk when selecting medications, giving consideration to dose and interactions between different pharmaceutical products. Likewise, patients have the responsibility to follow the instructions provided by their doctors for taking medications. Insomnia, often linked with anxiety, is a common complaint among the adult population and one of the reasons for the prescribing of, for example, benzodiazepines. Unwanted effects from such medicines include drowsiness, dizziness, unsteadiness, and blurred vision — all problems that are undesirable from a fall prevention perspective. Trials of behavioural sleep management programmes have found these to be effective in improving sleep and offer a viable alternative to use of hypnotic drugs. The effects of alcohol on coordination and balance are well known. Use of alcohol should be discouraged where fall hazards are present, and it is important to avoid the presence of fall hazards in locations where alcohol is consumed regularly (e.g., in bars and clubs).

Use of footwear commensurate with the prevailing underfoot conditions is a measure that can help everyone, from the wearing of suitable footwear for slippery outdoor conditions, through to use of shoes or boots with specialist soling in occupational situations where floor contamination cannot be avoided. Because the risk of falling is reduced if people can see where they are going, it is appropriate to promote regular vision testing, along with encouragement to use spectacles when needed.

12.5 Concluding Remarks

It is axiomatic that falls are as much a part of everyday life as births, deaths, and marriages. The complexity of the environments we negotiate on a daily basis, coupled with the fallibility of our species, means that it is inevitable that falls will continue to occur. Nevertheless, many falls are preventable. One important route toward addressing the problem is by raising awareness and encouraging attention to the falls problem, in the same way that safety on the roads is discussed frequently. It is hoped that this book makes a modest contribution in this respect.

Index